MICHAEL ROGOL WITH SUSAN HANEMANN ROGOL

RESHAPING ENERGY *and*
INFRASTRUCTURE *in* OUR
NEXT 1,000 DAYS

EXPLOSIVE
GROWTH

LIVE OAK
BOOK COMPANY

This publication is designed to provide accurate and authoritative information in regard to the subject matter covered. It is sold with the understanding that the publisher and author are not engaged in rendering legal, accounting, or other professional services. If legal advice or other expert assistance is required, the services of a competent professional should be sought.

Published by Live Oak Book Company
Austin, TX
www.liveoakbookcompany.com

Distributed by Live Oak Book Company
For information about special or bulk purchases, please contact PHOTON Consulting sales at explosivegrowth@photonconsulting.com or +1 617 904 0283.

Design and composition by Greenleaf Book Group LLC
Cover design by Greenleaf Book Group

Publisher's Cataloging-in-Publication data
Rogol, Michael.
    Explosive growth : reshaping energy and infrastructure in our next 1,000 days / Michael Rogol with Susan Hanemann Rogol.
    p. cm.
    ISBN 978-1-936909-32-2
    ISBN 978-1-936909-33-9 (e-book)
1. Solar energy industries --Forecasting. 2. Energy industries --Forecasting. 3. Electric utilities --Forecasting. 4. Power resources. I. Explosive growth : reshaping energy and infrastructure in our next one thousand days. II. Rogol, Susan Hanemann.
HD9502.A2 .R64 2011
333.79/12 –dc22   2011942380

Print ISBN: 978-1-936909-32-2
eBook ISBN: 978-1-936909-33-9

First Edition

FOR OUR PARENTS DIANE, ALAN, BARBARA, TOM

Who taught us that life is not a dress rehearsal

CONTENTS

KEY QUOTES—A QUICK OVERVIEW OF THE BOOK'S MAIN POINTS

- If Edison were alive, he would surely already be dressed in overalls and urgently working to reshape our energy and other infrastructure sectors over the next 1,000 days.

- The primary purpose of this book is to share with a broad audience what my team and I observe every day as we measure the explosive growth of new energy markets and the impacts on traditional energy markets.

- Twenty years into my career in the energy sector, I see strong evidence that major changes are happening now. These changes are occurring faster than nearly anyone expects and creating more value than nearly anyone would imagine.

- How fast? In the next 1,000 days, it is plausible that the fundamental economic and operational underpinnings of our energy sector may be completely reshaped.

- The driving force behind this reshaping will not be "peak oil" or "climate change" or other buzzwords on which the popular media often focus. Instead, the driving force for massive change in the energy sector will be "negative network effects."

- This book is written to enable CEOs and business executives to look more carefully at the opportunities and risks created by negative network effects.

- The core message is that negative network effects appear to be on a course to destroy the economics and operations of our energy infrastructure within 1,000 days.

- This is *not a prediction* that a rapid rise in the cost of our energy infrastructure and rapid shift to new energy technologies is occurring at a rate that will break traditional energy infrastructure. This is *an observation* in specific markets.

- It is plausible that volume, price and cost of traditional electricity companies and infrastructure will come under attack from solar power much like the USPS is under attack from online bill pay.

- These are the ten symptoms of a negative network effect. If you are observing them in your industry, you should go to the hospital emergency room. It will be the same story, only in a different industry setting.

- When negative network effects begin, they look small, expensive and linear. It takes a long time for a new technology to reach a cost structure and volume that is capable of disrupting the existing network infrastructure. It takes even longer to realize that the growth is exponential, not linear.

- My hope is that leading CEOs will aggressively seek paths to a heroic slaying or at least taming of the monstrous impacts of negative network effects.

- This hope is based on the strong conviction that solving these problems can be very profitable for companies and create value on a scale similar to Rockefeller and Carnegie.

- My gut says that a new era is being born, even if most people do not recognize it yet. This is the era of "distributed infrastructure industries."

- This movement creates potentially huge business opportunities.

- So this is my task: To convince CEOs that there is significant value from negative network effects.

- Negative network effects create massive, positive opportunities.

- Companies are well positioned with 1,000 day plans that have clear answers to the questions, "Where is value?" and "How to capture it?"

- No single company embodies being on the positive side of a negative network effect more than the leading network equipment vendor in the solar power sector, SMA.

- This company's performance has been outstanding. In 2009, during the heart of a global recession, the company grew revenue 37% year-on-year and achieved 24% operating profit margins. In 2010, SMA nearly doubled revenue and delivered 27% operating profit margins.

- It seems far-fetched, doesn't it? How the heck did this happen? How did SMA establish itself as the "Cisco of the solar power sector"? The simple answer is: SMA understands network effects. It knows how to target customers on existing networks who purchase high-price electricity and have a strong economic reason to switch to solar power.

- In 20 years of experience in the energy sector, I have not observed companies lacking real strength in energy acquiring energy companies such as Cypress Semiconductor's acquisition of SunPower and MEMC's acquisition of SunEdison.

- Similarly, in nearly two decades of experience working with Asian conglomerates, I have never observed a foreign acquisition that is so far from the core business as Sharp's acquisition of Recurrent.

- These are strange moves that indicate both greed and fear are involved.

- Why did these companies make these moves? These are not business-as-usual moves. Jumping from semiconductor/electronics technology and manufacturing into energy services and financing is not normal.

- This perspective is important for CEOs pondering a world filled with negative network effects and asking the central questions, "Where is value?" and, "How to capture it?"

- My hope is that CEOs will follow Rockefeller's advice about not being afraid to "give up the good to go for the great."

Introduction

THE NEXT 1,000 DAYS

Opportunity is missed by most people because
it is dressed in overalls and looks like work.
—Thomas A. Edison (1847–1931)

1,000 DAYS

A decade ago, I was a taxi driver in Seoul. It was a hobby pursued while working in South Korea as a management consultant for McKinsey & Company. The idea was to explore "normal" life in the Korean capital. When I was preparing for the taxi license test, a Korean friend asked, "What would you do if a pregnant woman went into labor in the back of your taxi during rush hour?"

I imagined the scene without any medical training or equipment: A crammed highway, the laboring noises of a soon-to-be-mother, blankets on the roadside, sweat dripping off my forehead and a doctor's voice on the phone trying to convey instructions while providing calm. The situation, of course, never became reality during my stint as a part-time taxi driver, but these days I often think about this hypothetical question.

A similar "what if?" is now a daily part of my professional life. "What if the trends I am observing today really do give birth to a new energy industry?" Many of my clients are asking this question because there may be opportunities today for value creation on a scale commensurate with Rockefeller, Vanderbilt, Carnegie and Gates. I agree. Twenty years into my career in the energy sector, I see strong evidence that major changes are happening now. These changes are occurring faster than nearly anyone expects and creating more value than nearly anyone would imagine.

How fast? In the next 1,000 days, it is plausible that the fundamental economic and operational underpinnings of our energy sector may be completely reshaped, particularly in the world's wealthy zip codes. The driving force behind this reshaping will not be "peak oil" or "climate change" or other buzzwords on which the popular media often focus. Instead, the driving force for massive change in the energy sector will be "negative network effects."

Most people are familiar with positive network effects. A positive network effect occurs when a network (e.g., an electricity network, a computer network, a railroad network) experiences rapid growth in volume. As volume grows, fixed cost per unit drops quickly. Lower cost often leads to lower price, which attracts even more volume to the network and further reduces cost. This dynamic is a positive network effect. It is familiar to anyone who observed the transition from high-priced, clunky mobile phones in the 1980s to lower-priced, smaller mobile phones in the 1990s and even lower-priced, sleek smart phones in the 2000s. Positive network effects underpin the economics and the operations of our infrastructure in energy and other sectors.

POSITIVE NETWORK EFFECTS »
(SCHEMATIC OVERVIEW)

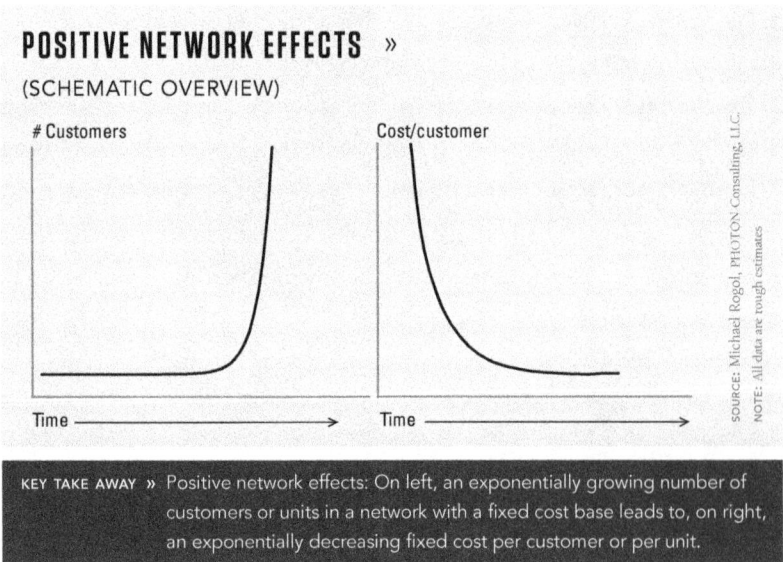

KEY TAKE AWAY » Positive network effects: On left, an exponentially growing number of customers or units in a network with a fixed cost base leads to, on right, an exponentially decreasing fixed cost per customer or per unit.

Most people do not yet realize that wealthy zip codes around the world are home to a dirty secret about energy: Network effects are running in reverse. The combination of (1) flat-to-declining population *and* (2) flat economic growth *and* (3) increasing energy efficiency for traditional consumers *and* (4) the rapid rise of new technologies that change energy consumption patterns mean that *many* of the world's wealthy zip codes are experiencing volume decline in traditional energy networks.

The result of these four factors is that fixed cost per unit is rising quickly, leading to higher prices. With higher prices, additional customers seek to be more energy efficient and also to use new technologies based on different energy inputs, which drives the cost higher still for traditional energy. This is a negative network effect.

My company, PHOTON Consulting, has conducted dozens of interviews with gasoline station operators in Connecticut. They talk about a flat economy plus a flat population plus more efficient traditional gasoline cars plus more hybrid cars leading to flat-to-declining volume through their stations. Similarly, electricity network operators in southern Germany (e.g., Bavaria) talk about a similar pattern at PHOTON's Solar Electric Utility conferences: Flat economy, flat population, more energy efficient households and more solar power systems.

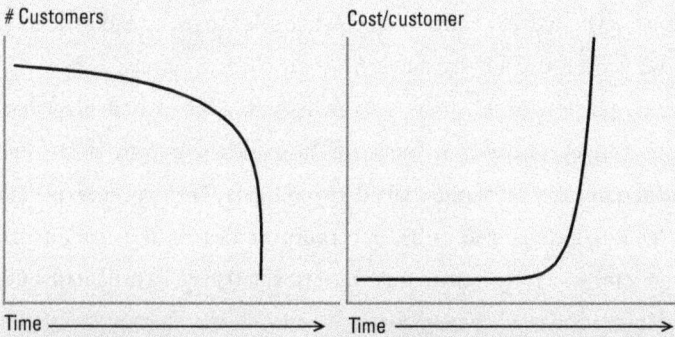

NEGATIVE NETWORK EFFECTS »
(SCHEMATIC OVERVIEW)

Customers

Cost/customer

Time

Time

SOURCE: Michael Rogol, PHOTON Consulting, LLC.
NOTE: All data are rough estimates

KEY TAKE AWAY » Negative network effects: On left, an exponentially decreasing number of customers or units in a network with a fixed cost leads to, on right, an exponentially increasing fixed cost per customer or per unit.

While a small subset of energy economists and energy executives study this trend closely, most people do not recognize that declining volumes on networks are *already* driving rapid increases in costs per unit. Similarly, most people do not realize that the operations of traditional energy networks are *already* broken in some markets. The trends that my team and I observe every day show that rising cost within our energy networks is happening much faster than nearly anyone anticipated. For example, electricity network operators and generators in southern Germany have higher costs from more manpower, more hardware and substations for substations, more hours of spinning reserve, less efficient fuel for traditional peaking units, more maintenance and other additional costs.[1] Further, we are watching as new energy technologies are being adopted much faster than anyone would have dreamed. As an example, global solar power supply has grown from 7 million watts per year in 1980 to 47 million watts per year in 1990 to 288 million watts per year in 2000 and 24,200 million watts per year in 2010.[2] Finally, we are seeing actual impacts on the economics and operations of traditional energy infrastructure that are much larger than anyone expected. For example, at the same time that costs are *up* in Germany's electricity market (e.g., in mid-2011, the price for natural gas is up 24% year-on-year and the price for coal is up 23% year-on-year), the price of electricity is *down* (in mid-2011, the price of wholesale electricity is down 1% year-on-year).[3] Our point is that fundamentals have *already* disconnected.

The primary purpose of this book is to share with a broad audience what my team and I observe every day as we measure the explosive growth of new energy markets and the impacts on traditional energy markets. The core message is that negative network effects appear to be on a course to destroy the economics and operations of our energy infrastructure within 1,000 days, particularly in the world's wealthy zip codes. In some places, it is plausible that this may occur within 1 year. This is not a *prediction* that a rapid rise in the cost of our energy infrastructure and rapid shift to new energy technologies is occurring at a rate that will break traditional energy infrastructure. This is an *observation* in specific markets.

Yet gloom-and-doom about problems in energy is *not* the reason for executives, investors and policy makers to pay attention. The reason to pay attention is the potential for massive value creation from the birth of new energy. Already in our sector there are more than 75,000 companies focusing on the positive sides of negative network effects. When cities (like Munich) host solar power conferences, they attract more attendees (80,000 at one recent conference) than many major sporting events with solar power companies occupying nearly all the cities' hotels and restaurants. These companies have generated more than $280 billion in revenue and more than $55 billion in operating profit over the last 5 years, with potential to continue growing quickly. And this is just within the energy sector. There is potential for a similar pattern to quickly emerge in the food, water and communications infrastructure of wealthy regions.

The implications are far-reaching. This book focuses much of its attention on the energy sector, but many other industries should pay careful attention. Specific industries that face game-changing shifts with massive potential upside and downside include:

- Agriculture
- Banking
- Chemicals
- Consumer products
- Electronics
- Executive search
- Food and beverage
- Health care
- Manufacturing
- Private equity
- Real estate
- Retail & wholesale
- Services
- Telecommunications
- Water

- Automotive
- Brokerage
- Consulting
- Department stores
- Energy
- Financial services
- Grocery
- Investment banking
- Pension funds
- Publishing
- Restaurants
- Securities & commodity exchanges
- Technology
- Transportation
- Venture capital

The last time the wealthy world underwent a rapid change in energy was 1860–1910, especially 1880 to 1900. During this period, Rockefeller's Standard Oil came to dominate the transportation side of energy while Edison and

Westinghouse came to dominate the electricity side of energy. The key decisions that drove the fortunes of Rockefeller, Westinghouse and Edison occurred early in the transition to the "new" energy of oil and electricity. At that time, the executives and companies that shaped our energy for a century and a half were not afraid to roll up their sleeves and get to work on hard problems.

Today, a new set of companies are making decisions that have potential to reshape energy within 3 years and influence the trajectory of energy for many decades to come. As the world awakens to the implications of negative network effects, the next 1,000 days are likely to be a period of major change as the birth of new energy creates massive wealth while also undercutting existing infrastructure in the wealthy world. The companies, executives, investors and policy makers that ignore negative network effects may destroy trillions of dollars in the value of our existing energy infrastructure. Conversely, the companies, executives, investors and policy makers that quickly focus on the positive sides of negative network effects have potential to create trillions of dollars in value.

This book is written to enable CEOs and business executives to look more carefully at the opportunities and risks created by negative network effects and to ponder the following questions:

- "What would you do if negative network effects drive the birth of new energy while also destroying old infrastructure?"

- "What if energy was just one among several sectors to go through a period of explosive growth and infrastructure transformation?"

- "And what if not being ready for explosive growth—being on the wrong side of negative network effects—had massive negative implications for you and your company?"

If Edison were alive, he would surely already be dressed in overalls and urgently working to reshape our energy and other infrastructure sectors over the next 1,000 days.

CHAPTER 1

A PERIOD OF CONSEQUENCES: MAIL DELIVERY NETWORK

The era of procrastination, of half-measures, of soothing and baffling expedients, of delays, is coming to a close. In its place we are entering a period of consequences . . .
—Winston Churchill (1874–1965)

CONSEQUENCES

The U.S. Postal Service (USPS) has entered a period of consequences driven by negative network effects. My hope is that CEOs will spend some of their immense mental energy thinking about the situation facing the postal communications network, pondering similarities/differences with risks facing other infrastructure networks and considering how their companies might benefit from finding solutions for these infrastructure networks. If this were *only* about USPS, I wouldn't suggest CEOs focus here, but I believe understanding the situation at USPS will open their eyes to a pattern impacting many other industries and companies.

The situation facing USPS displays the downside impacts of negative network effects. According to the USPS' 2010 annual report, the Postal Service racked up its worst year in history, with net loss rising from $2.8 billion in 2008 to $3.8 billion in 2009 and then more than doubling to $8.5 billion in 2010. In addition, the USPS faces a $238 billion cash shortfall in the coming decade. That's a big number, on average equal to roughly $24 billion per year or $250 per taxpayer per year during the coming decade.[1] To be clear, this is the *loss* after collecting payments for delivery of mail. Let's come back to this $238 billion loss over the next decade and what it implies for all of us.

First, though, let's unpack the reasons that might explain why the USPS has racked up such large losses and faces an even larger gap in the years to come. There is plenty to unpack from this baggage. Typical reasons offered by analysts of and executives within USPS to explain the financial problems include a weak economy, declining mail volume, prices far below international norms, unprofitable branches that are difficult to close for political reasons, competition from private sector firms such as FedEx and UPS for higher profit routes and products, limited flexibility to reduce the workforce and expensive retiree health benefits.[2] Overall, the USPS faces a mountain of problems.

This laundry list of problems is somewhat boring because most items have existed on a description of, "What's wrong with the USPS?" since the early 1990s, when I read case studies about USPS in business school. Yawn. Literally every item on the list in the preceding paragraph could have been included in a description of the USPS during the economic downturn of 2001–2003. Basically, this list of problems doesn't provide a compelling explanation for, "Why is USPS having much larger problems now compared to the past?"

This list is interesting for what is *missing*: Email. When email use expanded rapidly during the 1990s, USPS volume went *up* from 191 billion pieces in 1997 to 197 billion in 1998, 202 billion in 1999 and 208 billion in 2000.[3] Isn't that surprising? People had access to faster and cheaper ways to deliver written communication, but physical mail delivery continued to rise despite email. And volume continued rising after the economic slowdown in the early 2000s to reach 213 billion pieces in 2006. Isn't it counterintuitive that more physical mail went through the USPS during the decade that email penetrated our lives so thoroughly that *You've Got Mail* became a movie? Weren't we pecking away so much that "carpal tunnel syndrome" became a familiar ailment and AOL bought Time? Whereas the Buggles sang the truth at the birth of MTV with "Video Killed the Radio Star," the pattern does not apply to mail service. To paraphrase, "Email did *not* kill the postman."

So what is killing the USPS? What is different now compared to the 1990s through the mid-2000s? Why is the USPS talking about such large multi-year losses for the first time in its history? The "killer app" wasn't email or the Internet or text messaging or mobile phones or FedEx/UPS. The "murder weapon" was online bill pay. Paying bills online grew rapidly through the mid-2000s and began having a noticeable impact on the USPS by 2007. During that period, Harris Interactive and The Marketing Workshop ran a survey on consumer bill payment.

This survey showed steady growth in the number of online households and the share of online households that paid bills online. In their words,

"The 2007 Consumer Bill Payment Survey showed that, for the first time, online bill payments exceeded bill payments made by paper check among online households. Online payments made up 39 percent of the total volume of bill payments among online households, an increase of 4 percent over the previous December 2005 survey. In contrast, the volume of checks sent through the mail fell 4 percent to 34 percent of the overall volume."[4]

The smart people at Harris Interactive and The Marketing Workshop were witnessing the early stages of a vampire attack. Bill pay was sucking blood from the USPS. As online bill pay bit into the USPS' customer base, every drop of blood paid online was a drop not flowing through the veins of the USPS' network. From the perspective of the USPS, perhaps there was only one thing to say about online bill pay: "It sucks."

Online bill pay has particularly damaging characteristics from the perspective of an organization like the USPS. Most directly, every bill paid online reduces the volume of payments sent from customers who have received a bill in the mail. This is a one-for-one trade. If the vampire takes an ounce of your blood, you lose an ounce of blood. If you receive a single bill in the mail and pay that single bill online, the volume of mail through the USPS falls by exactly one. This would be a zero-sum game with the online bill pay provider winning volume and the USPS losing volume. Yes, it is as possible to generate new mail volume as it is to generate new blood, but this only works if you are talking about a small amount of blood relative to the 5 liters in your body.

Unfortunately for the USPS, this was not a small amount of blood because losing mail to online bill pay is worse than a zero-sum game when measured in profit terms. It isn't just that they lose a piece of mail. It's that they lose an above-average price and above-average profit piece of mail. The users of online bill pay are, by definition, online. Because household connectivity is more likely at the higher end of the socioeconomic range, this means that a bill paid online is more likely to be from a richer USPS customer than from a poorer USPS customer. On average, these customers tend to use higher-end USPS services that have a higher price and much higher margin for the USPS.

When these customers pay a bill online instead of through the mail, they abandon use of a high-profit USPS product, normally First-Class mail. In the words of USPS, "The decline in First-Class Mail . . . is especially disturbing as First-Class Mail remains our most profitable product. To compensate for the financial loss of the contribution of one piece of First-Class Mail, Standard Mail must increase by three pieces."[5] So online bill pay (by the virtue of the fact that it was available to higher-end customers who typically use higher-profit USPS products) is a targeted attack on the USPS' highest margin customers and products. The loss of a 3X-more-profitable piece of First-Class mail makes a single bill paid online much worse than a zero-sum game for the USPS. Online bill pay has a disproportionate impact on USPS profit, like sucking the nutrients from blood.

RELATIVE PROFIT IMPACT »

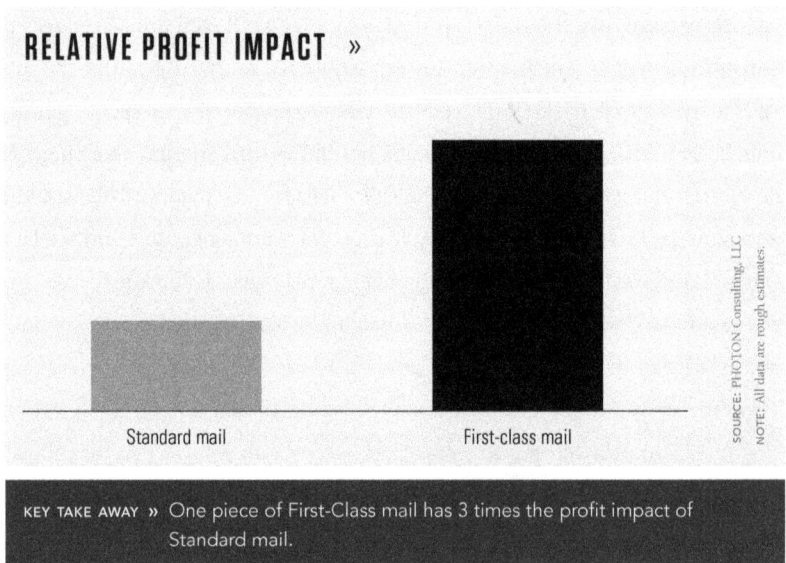

Standard mail   First-class mail

SOURCE: PHOTON Consulting, LLC
NOTE: All data are rough estimates.

KEY TAKE AWAY » One piece of First-Class mail has 3 times the profit impact of Standard mail.

But the payment of a single bill online is only the initial bite in a relentless attack. The heart of the problem is a multiplier effect. Customers who sign up to pay bills online typically pay *many* bills online during the course of a year. A survey of analyst reports covering online bill pay suggests that an upper-middle class household (a household that is among the most profitable for the USPS to serve) paying bills online typically pays ~15 bills online each month.[6] Because they do this while also shifting in many cases to online billing (in addition to only paying), they also eliminate bills arriving to them via the USPS. Further, marketing materials (marketing offers sent by a bill sender to a bill payer that are sent as separate letters) reaching these customers may also decline as the number of bills decline.[7]

This is a big revenue loss for the USPS for each upper-middle class household that adopts online bill pay. A reasonable estimate is that a typical upper-middle class household that converts to online bill pay eliminates roughly 300 letters per year from the USPS. This is ~15 checks per month sent from the customer via the USPS, ~10 bills per month sent to the customer via the USPS and an additional ~3 letters per month of marketing materials sent to the customer by various companies attempting to reach a potentially lucrative segment. This is ~28 letters per month, equivalent to ~330 letters per year. Assuming a USPS average revenue of $0.39/unit in 2010[8], this equates to a reduction of roughly $130 per year for each customer that switches to online bill pay.[9] That's a lot of biting into the body of USPS.

And the biting doesn't stop. What makes the loss of a customer to online bill pay particularly painful is the fact that most online bill customers remain online bill customers forever more. It's like a second marriage that leaves the first spouse to fall into destitution, homelessness and depression while the newly married couple lives happily ever after without concern for the former spouse. Once customers start paying a significant volume of bills online, they are likely to continue paying their bills online for many years to come. This is because banks have done an impressive job keeping their online bill pay customers satisfied.

One testament to the high level of customer satisfaction is the fact that households using online bill pay are significantly less likely (76% less likely according to one study) to churn away from a financial institution than households not using online bill pay.[10] In other words, online bill pay is "sticky" and keeps online bill pay customers with a specific financial institution. Great for the online bill pay company, but painful for USPS. The blood of the postman flows from a severe wound that bleeds year after year after year because the financial institutions have effectively locked in customers forever.

What makes this blood loss fatal for the USPS is not a customer paying a single bill online or a high-profit customer paying a single bill online or even a high-profit customer paying all bills online for all future years while also switching largely to electronic billing. The weapon used to murder the USPS (insert scary music crescendo of a horror movie attack scene) is the scale at which online bill pay and other forms of electronic payments suck away volume. So far, Bank of America, Wells Fargo and other U.S. financial institutions have drawn more than 30 million households into online bill pay.[11] This is just online bill pay, not including other forms of electronic payment. And this is just households, not businesses. To put this in context, 30 million check writers is equivalent roughly to one-third of the 90 million taxpayers in the U.S. who write a check to the U.S. government each year.[12] The overlap of online bill pay customers is not a perfect match with the numbers of taxpayers, but you get the point: 30 million online bill customers is a big number relative to the number of people writing checks to pay most of American private and public bills.

The scale at which customers convert to online bill pay is a sign that Bank of America (BoA; the largest online bill pay institution) and other financial institutions have a strong value proposition for customers. These 30 million customers already had a low cost, reliable, familiar pattern of paying bills through the USPS. They were drawn away from the reliable comfort of paying bills via the mail by a bill paying process that was cheaper, faster and easier. For many customers, online bill pay is a better solution. Overall, BoA has done an admirable job from the point of view of BoA and from the point of view of USPS customers.

The problem is *not* BoA offering online bill pay or millions of customers adopting online bill pay. The problem *is* the USPS network's dependency on growth. This is a life-and-death point. As long as a network with a fixed cost structure (such as the USPS) has a growing number of units, it is much easier for the network owner to capture profit by spreading its fixed cost structure over a growing volume of units. For many years, the USPS had experienced growing volume (e.g., as mentioned earlier, volume grew from 166 billion pieces in 1991 to 213 billion pieces in 2006). Over this period, operating revenue increased *every* year and operating revenue exceeded operating costs *every* year. In other words, the USPS showed operating profit *every* year from 1991 through 2006. While there are many sub-stories (e.g., effect of eBay and amazon.com on package deliveries), the main story that underpinned the USPS' long string of profitable years was growth. Volume growth made it easier for the USPS to increase revenue and maintain profitable operations.

When volume declines, though, networks with high fixed costs have fatal problems. This appears to be the case for the USPS. After 2006, the good times ended for USPS with volume declining from 213 billion pieces in 2006 to 212 billion in 2007, 203 billion in 2008, 177 billion in 2009 and 171 billion in 2010. It is important to note that the decline began *before* the economic collapse of 2008. The decline in USPS volume was *already* under way well before the collapse of Lehman in September 2008.[13]

Yes, the USPS volume was reduced by the macro-economic recession just as volume declined in car sales, airline travel and nearly every other sector. However, car sales recovered, airline travel rebounded and many other sectors bounced off 2009 lows in 2010. This story line does not apply to the USPS because mail volume is being dragged lower and lower by factors beyond the macro-economy, most notably online bill pay. USPS volume fell by 42 billion pieces between 2006 and 2010 and from 213 billion pieces in 2006 to 171 billion in 2010. Of this 42-billion piece reduction in annual mail volume, a "best guess" is that online bill pay single-handedly reduced USPS volume by 10 to 15 billion pieces of mail per year as of 2010.[14] In other words, online bill pay is responsible for approximately a quarter to a third of the lower annual mail volume since 2006. Given this trend, it appears that online bill pay was responsible for ~$5–$7 billion in lost annual revenue in 2010. In the context of USPS' $8.5 billion loss in 2010, the majority can be attributed to online bill pay.

Wow. This is surprising. And I suspect it will be surprising to many that a big chunk of the USPS' problem is directly attributable to volume, revenue and profit lost to online bill pay. Many analysts discuss the impacts of slow economic and population growth, substitution by email, Internet and mobile telecom and competition from FedEx and UPS. All of these are standard-issue explanations for what is wrong with the USPS. And all likely *do* have some part to play in the woes facing the USPS. But the single biggest factor driving declining USPS volume is almost certainly online bill pay.

My hope is that this situation will grab the attention of CEOs and other executives enough for them to think more about the problem facing the USPS, how to solve it and how solutions to the USPS problem might apply to other similar problems (i.e., pattern recognition). It's important that smart businesspeople

USPS MAIL VOLUME »
(BILLION PIECES/YEAR)

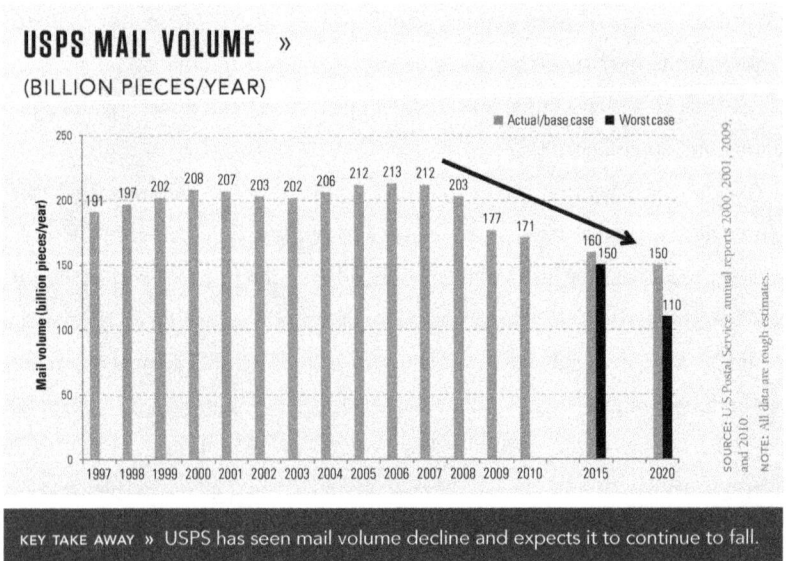

Legend: ■ Actual/base case ■ Worst case

Data (billion pieces/year):
- 1997: 191
- 1998: 197
- 1999: 202
- 2000: 208
- 2001: 207
- 2002: 203
- 2003: 202
- 2004: 206
- 2005: 212
- 2006: 213
- 2007: 212
- 2008: 203
- 2009: 177
- 2010: 171
- 2015: 160 (actual/base case), 150 (worst case)
- 2020: 150 (actual/base case), 110 (worst case)

SOURCE: U.S. Postal Service annual reports 2000, 2001, 2009, and 2010. NOTE: All data are rough estimates.

KEY TAKE AWAY » USPS has seen mail volume decline and expects it to continue to fall.

think about the USPS problem more because the problem looks like it will only get worse (hence solving the problem will only get more valuable). Nearly all analysts covering online bill pay expect continuing growth. For example, Forrester expects U.S. online bill payment households to increase from 48 million to 66 million by 2014.[15] That's an increase of ~40% in the number of households by 2015 before accounting for the additional impacts of increases in the number of online bills paid per customer. Online bill pay *alone* might reduce USPS volume by ~5 billion pieces per year by 2015.

The USPS knows it has a problem. Its own estimates on page 3 of the *USPS 2010 Annual Report* are for total mail volume to decline another ~10–20 billion pieces to ~150–160 billion pieces by 2015. In other words, the ~5 billion pieces of USPS volume lost to online bill pay account for ~25%–50% of the total reduced mail volume expected by the USPS through 2015. Consultants to the USPS echo this view, including Boston Consulting Group's analysis "Projecting U.S. Mail Volumes to 2020," which identifies nine main drivers of declining USPS mail volume in the coming decade. The point is not that online bill pay is the only problem facing the USPS, but it is a major piece of the problem and is worth understanding

USPS OPERATING REVENUE, EXPENSES AND PROFIT/LOSS »
($ BILLION)

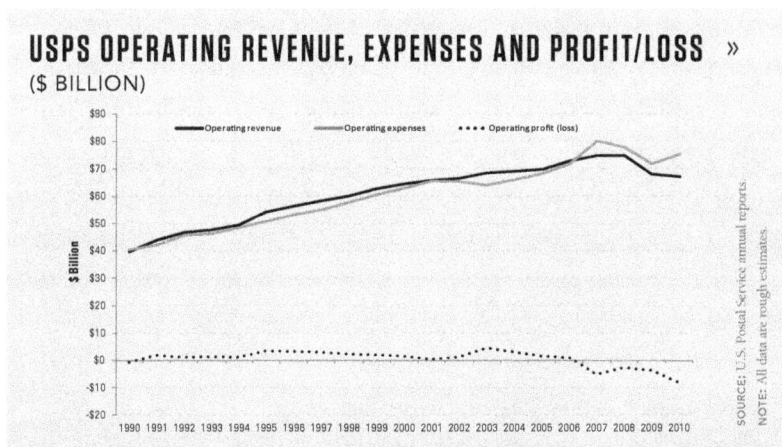

KEY TAKE AWAY » Volume decline has driven USPS operating losses since 2007.

because it is a good example of the type of technology transition that causes problems for existing infrastructure networks.

The core challenge for the USPS is that its mail delivery network has a high fixed cost structure that cannot easily be reduced. It is required to keep most of its offices open, continue delivering in areas even if there is little volume and continue service six days per week. These requirements set a high level of cost for the USPS that must be paid regardless of the volume of mail through the USPS network. The key point is that the USPS (like many other infrastructure networks) has high fixed costs/low variable costs with pricing that is predominantly variable. When there is a decline in the number of units processed within a network that has high fixed costs, the fixed costs per unit increases. If the number of units declines quickly, then fixed cost per unit rises quickly. This is the problem facing the USPS, a problem being driven by online bill pay, among other factors.

It would be a mistake to think that the challenge of online bill pay is *just* a company-level problem for the USPS. It goes beyond the company level. Don't mistake the tree for the forest. This is a problem that undermines the economics of the U.S. mail infrastructure network, not just the economics of the company responsible for operating the network. This distinction is important. The volume on the existing U.S. telephone network increased during the 1980s and 1990s even while many people switched to using mobile telecom and the Internet. The problem was that AT&T and other traditional telecom companies had business models that were unable to compete in an era of new telecom technologies. Many traditional telecom companies went bankrupt. However (and this is the crucial distinction), the traditional telecom networks continued to have increasing volume of use, albeit for new reasons (more data volume, more volume connecting mobile phones to landlines, etc.). As a result, traditional telecom companies died. However, the traditional telecom infrastructure networks live, but with changes in ownership of traditional telecom networks.

The problem for the USPS is *both* at the company level and at the network level. *Declining* volume through its network means that *both* the company and the network are in a death spiral. This is the core of a negative network effect. It is bigger than a no-longer-effective business model and bigger than "new technology replacing old technology" because it is about a change of business model, plus a change of technology, plus a change of network. The old company, technology *and* network get replaced as the result of a negative network effect. Because this is an infrastructure network that may fail, it is not just the company involved that loses. There may be significant collateral damage for other businesses and other people who rely on the traditional infrastructure. But this gets ahead of our story. More on this later. The characteristics of negative network effects (shown in the following figure) are applicable to more case studies that we will discuss throughout the book.

For now, the main point for CEOs and other executives trying to create value is that the USPS serves as an example of a negative network effect in an infrastructure industry. Within this example of a negative network effect, it is worth focusing on ten key characteristics of the USPS' situation:

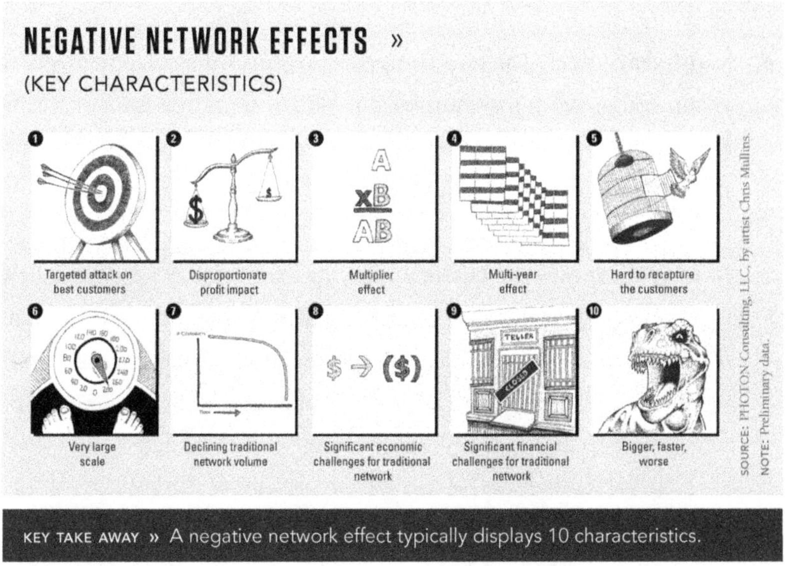

NEGATIVE NETWORK EFFECTS »
(KEY CHARACTERISTICS)

❶ Targeted attack on best customers	❷ Disproportionate profit impact	❸ Multiplier effect	❹ Multi-year effect	❺ Hard to recapture the customers
❻ Very large scale	❼ Declining traditional network volume	❽ Significant economic challenges for traditional network	❾ Significant financial challenges for traditional network	❿ Bigger, faster, worse

SOURCE: PHOTON Consulting, LLC, by artist Chris Mullins.
NOTE: Preliminary data.

KEY TAKE AWAY » A negative network effect typically displays 10 characteristics.

1. **Targeted attack on the best customers:** A new competitor uses a new technology to take volume away from an existing network by targeting customers that purchase the highest-priced products. In this case, BoA and other financial institutions used online bill pay to target USPS' First-Class mail customers.

2. **Disproportionate profit impact:** Because the volume removed from the network is high price, the impact on profit for the existing network owner is many times higher than the impact on volume. For the USPS, the profit impact of losing one First-Class mail letter is 3X the volume impact.

3. **Multiplier effect:** The new competitor removes the customer from the traditional network for many transactions, not just a single transaction. For the USPS, an upper-middle class household joining BoA bill pay reduces USPS volume by ~330 letters per year, equivalent to ~$130/year of revenue reduction.

4. **Multi-year effect:** The new competitor removes the customer from the traditional network for a sustained period of time, perhaps for decades. For the USPS, customers who join online bill pay demonstrate a propensity to remain with their financial institution for many years.

5. **Hard to recapture the customer:** Because the value proposition offered to the customer by the new competitor is so strong and has lock-in, it is difficult for the old network to recapture the customer. For the USPS, it is difficult to compete with the value proposition of BoA online bill pay, which is cheaper, faster and easier to use for many customers.

6. **Very large scale:** The new competitor is able to very quickly expand and capture a significant share of the old network's customers and volume, especially the old network's highest-profit customers and volume. For the USPS, there are already more than 30 million households in the U.S. that pay bills online.

7. **Declining traditional network volume:** The rapidly rising scale of the new competitor reduces the volume through the traditional network. For the USPS, volume declined by 42 billion pieces per year (–20%) from 2006 to 2010, with online bill pay alone responsible for 10 to 15 billion pieces per year of reduced USPS mail volume. The decline in volume through the USPS network is likely to continue in the coming years according to a broad set of experts.

8. **Significant economic challenge for traditional network:** Because the traditional network has high fixed costs (often financing costs or historical employee benefit costs), it faces a rapid deterioration in its economic performance, with volume and revenue declines not matched by similar cost declines, leading to losses. For USPS, online bill pay reduced USPS revenue by $5–$7 billion in 2010, accounted for a significant share of USPS' $8.5 billion net loss in 2010 and is poised to continue driving lower volume, reduced revenue and larger losses for the USPS in the years to come. It is important to note that this economic challenge is both for the company and for the underlying network that the company owns/operates.

9. **Significant financial challenge for traditional network:** With declining volume and revenue and profit, the traditional network finds it difficult to continue refinancing, making it impossible to invest into technologies, strategies, or business models. USPS, which receives its financing from the U.S. government, was added to the federal government's General Accounting Office's list of high-risk areas in July 2009, citing the USPS' expected inability to cover its financial obligations.[16] This situation is different than for traditional U.S. telecom networks that had rising volume in the 1990s and were readily financed by new owners with new business models without significant interruption of service for most customers.

10. **Bigger, faster, worse:** All of the this happens at a scale that is larger and at a pace that is faster than most outside observers anticipate, leading to

impacts that are worse than nearly anyone expects. For the USPS, admitting to a $238 billion cash shortfall in the coming decade seems to be a reasonable quantification of the potential losses coming for the entity through 2020, but there are no credible actions yet being taken at a scale sufficient to prevent these losses and their impacts on the broader society. In addition, unlike the telecom example in which there were obvious potential purchasers of the traditional telecom networks, there are not (at least not yet) obvious potential purchasers of the USPS network.

These are the ten symptoms of a negative network effect. If you are observing them in your industry, you should go to the hospital emergency room. It will be the same story, only in a different industry setting.

To me, point 10 is at the core of why I want senior business executives to think more about the USPS. If a thoughtful analysis shared by the USPS suggests that it is poised to lose $238 billion in the coming decade, it seems that the USPS is entering, in Churchill's words, a period of consequences. If so, what action could be taken to reduce this risk and eliminate this loss? Are there better business models and better owners for this network? The answer to this question is worth $250 per taxpayer per year for the next decade. I am hopeful that this is a big enough problem to capture a CEO's attention for at least a little while. I am also hopeful that if the CEO recognizes that this problem and its solution are applicable to a broad set of networks, that the CEO might just consider quickly shifting business resources into areas that are on the positive sides of negative network effects.

The mailman has continued delivering despite rain, sleet and snow. But now the USPS faces a bigger problem: Declining volume is driving a negative network effect. With an expectation that the number of units on its network will decline in the future, profit will become much more difficult and network owners, executives and financiers will run away as if fleeing a monster. There is no obvious stake or garlic to slay or repel the vampire. The central point is that networks are built on an economic foundation of volume growth. Without volume growth, USPS' network faces an economic horror movie. Unfortunately, this is just the first in a series of horror movies. Now let's watch another.

Chapter 2

INSANITY: ELECTRICITY NETWORK

Insanity: Doing the same thing over and over again and expecting different results.
—Albert Einstein (1879–1955)

INSANITY

CEOs from a broad set of industries have been invited to opening night of another economic horror movie. This movie is the sequel to the USPS film. Similar to the USPS movie reel, this next movie focuses on real-world events and makes painstaking efforts to be accurate.

The opening scene of the sequel horror film begins in 1976. On the screen is the image of a large, suburban home with solar power panels. These panels sit on the roof and generate electricity when the sun is shining. The solar electricity is used during the day. The house is also connected to the standard electricity infrastructure network and uses network electricity from the utility on cloudy days and at night. The sun is shining in the foreground, but music in the background is dark and ominous. We are looking at a haunted house.

SUBURBAN "HORROR HOUSE" WITH SOLAR POWER »
(ILLUSTRATIVE)

SOURCE: iStockphoto

KEY TAKE AWAY » Houses with solar power are a mystery with hidden meaning.

In 1976, the first year for which data are available about the supply of solar power, roughly 2 million watts of solar panels were manufactured in small batches by a cottage industry.[1] Most of these solar panels were installed in test facilities for research or in remote cabins without any access to electricity networks. The suburban house in our second movie was an enigma, a mystery with hidden meaning. It should not have existed with solar power. It was cheaper for the house to use electricity from the utility than to purchase the solar panels; so this house was either inhabited by a crazy owner, or there was something foul in the air. From an economist's point of view, this customer was making a bad economic decision. We hear a shrill scream and the scene cuts to black. The audience is left with the creeping sense that something bad is going to happen.

The next scene opens with me on the bullet train from Tokyo. It is 2002, two and a half decades later. I am on the way to Osaka for a meeting with executives at a large energy company as a management consultant with McKinsey. I am in a window seat on the north side of the train looking out at the towns as the train raced. The buildings are facing me, facing south. Looking out the window, there are a remarkable number of solar power panels on the roofs of houses and apartment buildings. I think, "Strange to see so many solar panels." So I start to count roofs with solar as we speed past. My guess is that 1%–2% of the buildings have solar power on the roof. This makes no sense. Solar power is the most expensive way ever invented to generate electricity. I cannot understand why there are so many solar panels. The sun is shining on the Japanese suburbs while the dark music in the background grows slightly louder.

Like many people who develop expertise in an industry, it is common to see something you don't know about, but rare to see something you fundamentally don't understand. By 2002, I am a dozen years into a career in the energy industry. I have a pretty solid foundation in energy, but I do not understand what enables those solar panels to be on so many roofs. A week later, I wiggle an introduction to the CEO of Sharp Solar, the subsidiary of the Japanese electronics maker that is the world's number one solar power panel manufacturer in the early 2000s. The meeting is brief but crucial for the plot of this movie. Attempting to be polite in a Japanese context, I ask Tomita-san, a conservatively dressed 50ish-year-old, silver haired executive first about his hometown and his family. Then we move to a discussion of the Japanese economy and society and then to prospects for Sharp and its LCD TV business. His commentary on all these topics: Flat . . . flat . . . flat . . . flat . . . flat. He says nothing energetic or optimistic. Everything he says suggests that the future would be an uninteresting continuation of the present—bored and boring.

Then I asked him, "What about the future of solar power?" The camera angle pulls in tight on his face. He leans toward me. His answer is much more forceful and totally different: "It will be a trillion-dollar industry by the time I retire." This seems absurd. In 2002, the solar power industry's 560 million watts of supply is 280X larger than the 2 million watts in 1976, but the revenue pool is still only $4 billion in 2002.[2] So reaching a trillion dollars would require another 250X growth in revenue. This would require 15 years at ~45% compound annual growth. No sector has ever grown that fast for that long.[3] My reaction is, "Impossible," but I am curious and puzzled. I can't understand why there are so many solar panels on Japanese roofs or why an otherwise demure Japanese business executive is making an outlandish forecast about solar power's growth. The scene closes with mystery in the air.

Flash forward another 5 years. It's 2007. My skin is white and pasty from 60 months locked in a PhD program at MIT's Laboratory for Energy & Environment. More than 1,800 days in a dungeon, with a single question dripping on my brain like water torture: "Why did the solar power sector develop so quickly in Japan?" No sleep, little food, only this relentless question. The answers I find are so startling that I abandon the PhD even after completing a 181-page master thesis and taking my PhD general exams. What I have learned haunts me. I feel similar to the little boy in the movie *The Sixth Sense* who says, "I see dead people," but no one believes him. The solar power sector keeps multiplying, and I keep warning of consequences, but I appear crazy to the audience.

On screen, I begin talking about how solar power could quickly drive "saturation" of traditional electricity networks. In 2007, I publish a brief overview of solar-driven saturation of traditional electricity infrastructure that may come by 2012, writing about the "inability for the electricity grid to operate with high reliability as more solar power is added." By 2008, the sector revenue pool reaches $43 billion, 10X larger than 2002 and equivalent to a 50% compound annual growth rate.[4] This growth rate is *faster* than the path to $1 trillion described by the Japanese executive in 2002. Music in the background mimics a heartbeat accelerating with fear.

The scene continues in 2008 with me warning of the implications for traditional electricity in a monologue about tragedy on the horizon:

> Perhaps the most important implication of solar's rapid volume growth is that it will quickly "converge" into traditional electricity markets, taking over a substantial portion of peak electricity consumption in several markets . . . In the coming years, solar power is poised to become a much larger piece of the broader electricity sector—and certainly larger and faster than traditional electricity companies expect . . . This is a positive story of solar's ascent and its emergence as a meaningful piece of electricity markets. Yet this positive story of solar's growth will create important risks for traditional electricity players. As more end-customers adopt solar power, many traditional electricity players will be faced with three important challenges. First, the adoption of solar by end-customers will require traditional electricity companies to identify alternative sources of income to make up for revenues displaced by solar power systems. Second, the rising scale of solar will require that traditional electricity players adapt their operations to account for the larger share of electricity contributed by photovoltaic (PV) systems. Third, because the scale and timing of solar adoption in any specific market are uncertain due to global market forces, traditional electricity players will need to address new forms of uncertainty and risk that will impact financing, among other effects.[5]

The words, "traditional electricity players will need to address new forms of uncertainty and risk that will impact financing" echo as the scene reaches its end with the audience tensing in anticipation. "This is just like the first horror movie before the mailman got crushed. Blood is about to flow."

And they are right. Cut to Spain at the end of 2008 to view an injured electricity sector sitting in a small pool of blood like a boxer on the mat after smelling salts awaken him from a knockout. Spain's traditional electricity sector has been aggressively attacked and maimed, though not killed. The Spanish solar power market grew from 10 million watts in 2004 to 20 million watts in 2005, 75 million in 2006, 480 million in 2007 and 2.5 billion (billion not million) in 2008.[6] Look at the pace of this attack—2X in 2005, another ~4X in 2006, another ~6X in 2007 and another ~5X in 2008! That's 250X in just 4 years. This rapid expansion saw solar power balloon from 0.02% of Spain's electricity generating capacity in 2004 to 3.5% in 2008.[7] In 2008 alone, this expansion creates a $2.2 billion annual gap in the Spanish electricity sector (compared to $0.1 billion expected prior to 2008) because solar power uptake occurred so much faster than the government expected.[8] The Spanish government quickly steps in with policies to limit the volume of new solar power installations, so that new Spanish solar power installations decline by 95% in 2009.

The audience watches the attack of solar power in Spain and sees negative impacts on Spain's traditional electricity network. Yet many are skeptical that this is truly an attack by a monster with potential to undercut the fundamental foundation of the electricity system. Many people in the audience think, "This isn't real. This is not another USPS. It's just special effects. In reality, the problem was just bad government policy."[9] This thinking is dead wrong. Yes, the government's solar power policy in Spain was poorly conceived, poorly executed and did not anticipate the rapid pace at which solar power would be adopted. However, this alone does not explain how $19.6 billion (again, billion not million) was spent on new solar power installations in Spain in just 2008. That's $425 per person for every one of the 46 million people in Spain.[10] It also does not explain how a gap of $2.2 billion was created in Spain's electricity sector in just 1 year.

SOLAR POWER INSTALLATIONS IN SPAIN »
(MILLION WATTS/YEAR)

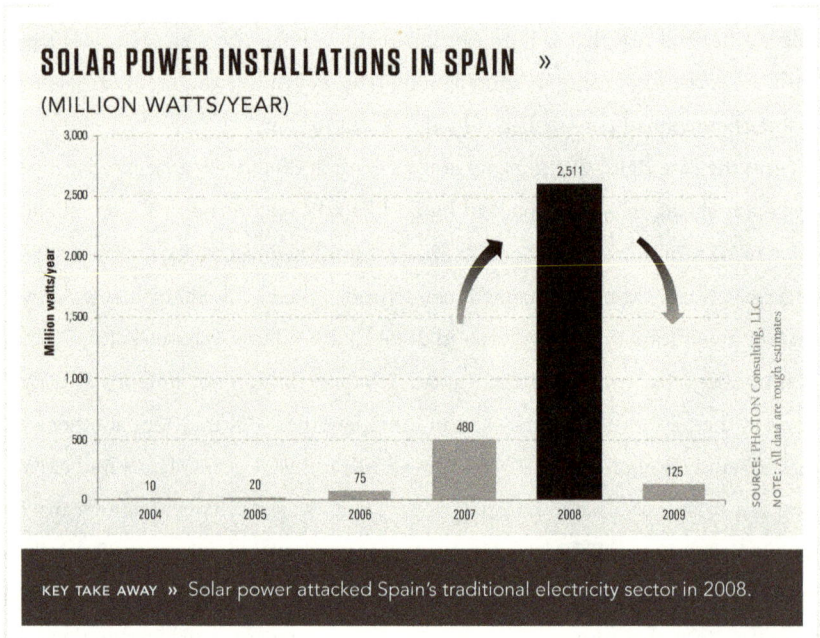

SOURCE: PHOTON Consulting, LLC.
NOTE: All data are rough estimates

KEY TAKE AWAY » Solar power attacked Spain's traditional electricity sector in 2008.

The core of the issue is that solar power attacks traditional electricity networks with incredible speed. The poorly designed policy enabled the monster to attack in Spain, but it should not distract anyone from the much more important point: A monster attacked traditional electricity infrastructure with blinding speed and painful financial impact. Paying attention to the policy and not the monster is like wearing your seatbelt while driving a car off a cliff.

Some people in the audience are genuinely scared, but many remain skeptical, pointing out that there are many examples of economic misadventure in Spain. So the horror film zips to Europe's largest economy, Germany, which becomes the world's largest solar power market after the Spanish attack ceases. In 2008, Germany installed 1.8 billion watts of solar power.[11] It might have been larger, but the solar market in Spain left few solar panels for the German market. When Spain caps its market at the end of 2008, the solar panels begin to attack Germany. Look at the results: Germany grows from 74 million watts in the first quarter of 2009 to 471 million in the second quarter of 2009, 926 million in the third quarter and 2.3 billion in the fourth quarter.[12] In other words, the fourth quarter of 2009 is 30 times larger than the first quarter of 2009. The monster is attacking Germany in 2009 after leaving Spain for dead at the end of 2008. And the attack has more force in 2009, with Germany installing virtually the same amount in just the fourth quarter alone that Spain installed during all of 2008 when it was under attack from the monster. The monster is getting stronger and the attacks more brazen.

SOLAR POWER INSTALLATIONS IN GERMANY »
(MILLION WATTS/QUARTER)

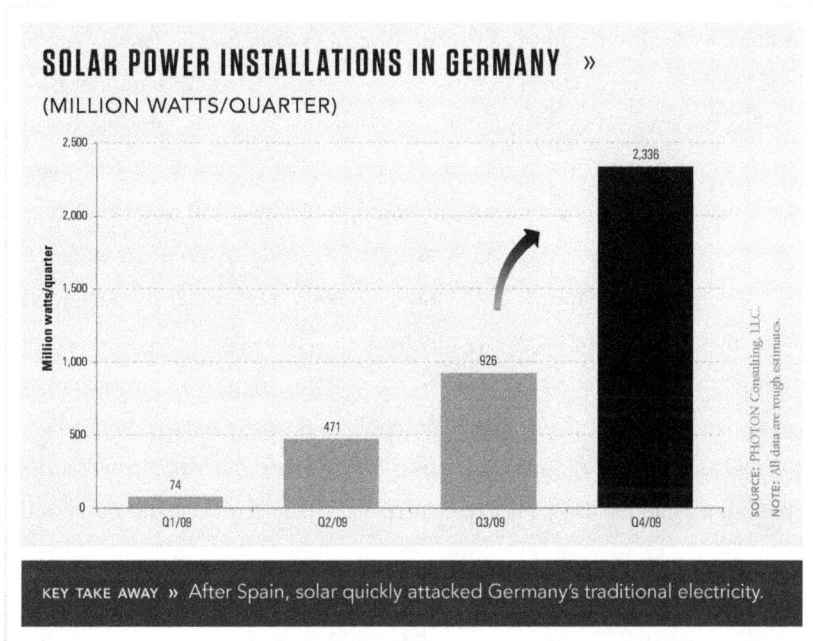

SOURCE: PHOTON Consulting, LLC.
NOTE: All data are rough estimates.

KEY TAKE AWAY » After Spain, solar quickly attacked Germany's traditional electricity.

These quarterly comparisons, of course, can be "explained away" as the result of "seasonal factors" (slow in winter, faster through the year). However, this is just wishful thinking instead of recognition that the monster appears real. Seasonal factors do not explain why Germany grew ~2X to 4 billion watts for the full year 2009 and another ~2X to 9 billion watts in 2010. This is significant growth on a fairly large base. Seasonal factors do not explain how 150,000 buildings in Germany added solar power systems to their roofs in 2009 and how this grew 2X in 2010. Seasonal factors do not explain how solar power capacity has quickly grown to be 14% of Germany's traditional electricity infrastructure with nearly none of that capacity owned by the traditional electricity companies.[13] Seasonal factors do not explain how the rapid addition of solar power systems on so many small buildings is causing noticeably lower prices in the German wholesale electricity market during the middle of the day when the sun is shining.

A member of the audience can try to explain the monster away, but even the headlines recognize the impact of solar power. For example, Bloomberg reports, "Solar Doubling Drives Down German Power Prices."[14] The idea that solar power is having negative impacts on traditional electricity should be obvious because there is a lot of solar power going into Germany very quickly. To put scale in perspective, Germany as a country has roughly $100 billion in revenue each year from all traditional electricity end-customers.[15] In comparison, these same end-customers spent an additional ~$28 billion on solar power installations (predominantly solar installations on small buildings) in 2010 alone.[16] That's a lot of solar power predominantly on small, distributed rooftops, equivalent to $345 per person for everyone in Germany, a sizeable number for such a sizeable population.[17]

With many people adopting solar power very quickly, it is obvious that there will be downside for the traditional electricity sector. A reasonable guess at the impact from a combination of lower volume, lower price and higher cost is a reduction in profit for Germany's traditional electricity sector by $5 billion to $10 billion in 2010.[18] This is a noticeable impact in a market with roughly $100 billion in end-customer revenue and roughly $10 billion in operating profit before taking solar power into account. Basically, solar power wipes out between half and all of the profit in Germany's traditional electricity sector in 2010. While these estimates are certainly worth debating,[19] the simple point is that the monster demonstrates faster pace and more painful impact in Germany in 2009–2010 than it had in Spain in 2007–2008. The organ continues an ominous tune.

The monster is gaining strength. This is demonstrated by other attacks occurring at the same time as the German attack. Whereas the attack in 2007–2008 was largely confined to Spain, there are attacks in 2009–2010 in Germany and many other locations. Cut to the Czech Republic with higher-pitched music in the background. Three million watts in 2007 rising to 51 million in 2008, 406 million in 2009 and 1.3 billion in 2010.[20] This is 433X growth in 4 years, faster than the solar power attacks in Spain or Germany. For a country with only 25% the population of Spain, this is twice as much volume installed per person in the Czech Republic in 2010 (130 watts per person) compared to Spain in 2008 (55 watts per person). In dollar terms, $5.6 billion was spent on solar power in Czech Republic in 2010 alone, equivalent to more than $500 per person for every one of the country's 10 million people.[21] By the end of 2010, solar power capacity in Czech Republic was roughly 9% of traditional electricity capacity, up from 0.02% in 2007.[22] The implication, according to Bloomberg *BusinessWeek*, is, "Czech Solar Power Risks Industrial Power Price Surge."[23]

Monstrous music reaches its final peak with monster attacks at a frenetic pace in France (6X from 2008 to 2010), Greece (11X from 2008 to 2010), Italy (9X

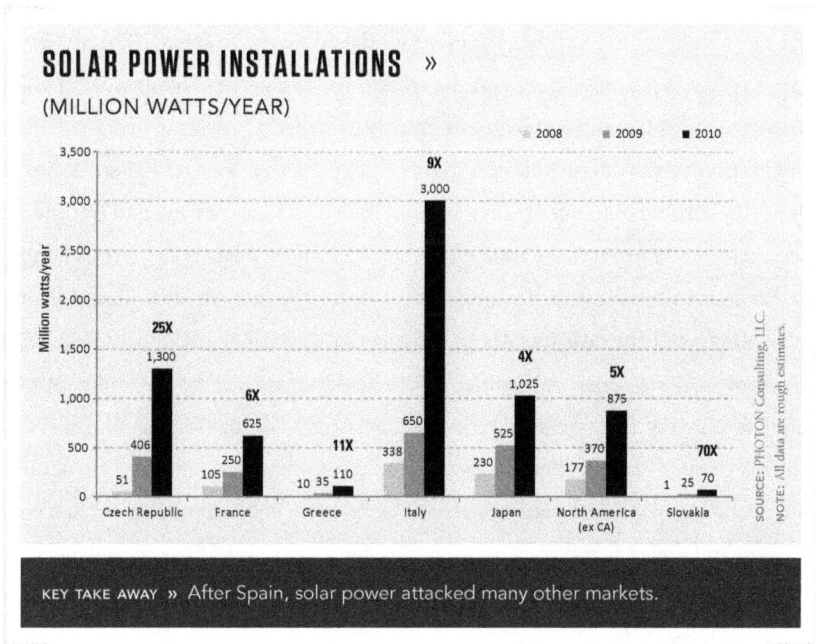

SOLAR POWER INSTALLATIONS »
(MILLION WATTS/YEAR)

KEY TAKE AWAY » After Spain, solar power attacked many other markets.

from 2008 to 2010), Japan (4X from 2008 to 2010), Ontario (3X from 2009 to 2010) and Slovakia (70X from 2008 to 2010).[24] The list goes on. With each attack, it becomes clear that the global economic slowdown has not diminished the monster's appetite. The global supply of solar power *accelerates* through the 2008–2010 economic downturn (+79% in 2008, +81% in 2009 and +76% in 2010), as do negative impacts on traditional electricity.[25] Fear grips the audience as the monster continues gaining strength and causing more bloodshed from the traditional electricity sectors across a broadening geographic footprint. Blood flow covers the screen red, and then it fades to black.

The movie's closing shot slowly comes into focus. It is early dawn on the front lawn of the haunted house from 1976. As the movie ends, the camera pans out to show another and then another and then many other similar houses with solar power on the roof. In 2009, there were more than half a million small buildings like this around the world that installed new rooftop solar power systems just from January through December. In 2010, the number of new solar power rooftops rose to more than 1 million per year. This final shot fades to black as more and more solar power buildings cram to fill the space of the screen. The audience full of CEOs is left with the haunting question, "What does the growth of solar power mean for traditional electricity infrastructure?"

The honest answer to this question is, "No one knows." There are a broad range of plausible outcomes from the traditional electricity players and traditional electricity infrastructure. Anyone who claims to know what is in store is simply not familiar with the massive complexity of electricity markets. Is it plausible that solar power will hit new hurdles to growth that it has not previously faced and stop growing? Yes, this is plausible. Is it plausible that solar power will continue to grow at a very fast rate and traditional electricity companies will find ways to address many of the challenges created for traditional electricity infrastructure? Yes, this is plausible. Is it plausible that solar power will continue to grow at a very fast rate and create negative network effects for a broad set of traditional electricity companies and traditional electricity infrastructure? Yes, this too is plausible. There are many "plausibilities."

GLOBAL SOLAR POWER MANUFACTURING OUTPUT »
(ANNUAL SUPPLY IN MILLIONS OF WATTS)

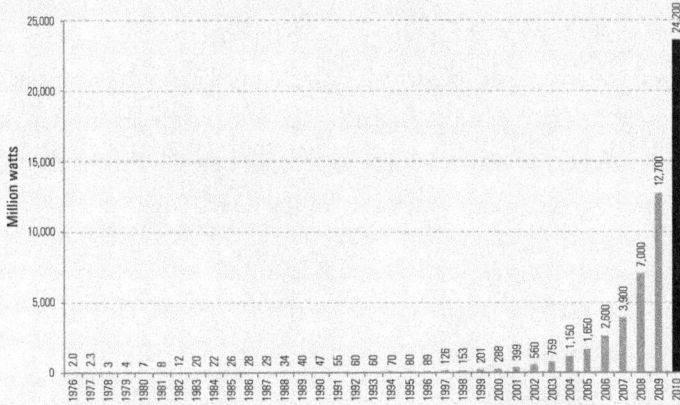

KEY TAKE AWAY » Relentless growth of solar power supply since the 1970s.

While there is no way to predict, it seems most reasonable to pay atten-tion to Einstein's definition of idiocy: "Doing the same thing over and over again and expecting different results." In this case, we have observed over and over again:

- Solar power supply growing relentlessly for 35 years straight;

- Solar power installations growing very quickly in a broad set of markets; and

- Negative impacts for traditional electricity companies and infrastructure in markets with high penetration of solar power installations.

Given that the price of solar power remains well above the cost for most manu-facturers (more on this later in the book), it seems plausible (not definite but plausible) that the results will continue to be the same (i.e., strong supply growth will continue). Given that the cost of solar electricity from small building roof installations is well below the price of traditional grid-based electricity in a broad portion of the wealthy world, it appears plausible (not definite but plausible) that the results will continue to be the same (i.e., rapid uptake of solar power in end markets). Given the rapid growth of solar power in specific markets, it appears plausible (not definite but plausible) that there will be significant challenges for traditional electricity (i.e., negative impacts for traditional electricity players and infrastructure).

NEGATIVE NETWORK EFFECTS »

(KEY CHARACTERISTICS)

❶ Targeted attack on best customers	❷ Disproportionate profit impact	❸ Multiplier effect	❹ Multi-year effect	❺ Hard to recapture the customers
❻ Very large scale	❼ Declining traditional network volume	❽ Significant economic challenges for traditional network	❾ Significant financial challenges for traditional network	❿ Bigger, faster, worse

SOURCE: PHOTON Consulting, LLC, by artist Chris Mullins. NOTE: Preliminary data

KEY TAKE AWAY » Declining traditional network volume (#7) is among the most important characteristics of a negative network effect.

Given this plausibility, it makes sense to evaluate the potential for traditional electricity infrastructure to face a negative network effect caused by the rapid growth of rooftop solar power. The pattern of a negative network effect appears to be similar to the first USPS movie:

1. **Targeted attack on the best customers:** Solar power typically attacks end-customers with the highest electricity price. These customers are usually small building electricity users with high rates in wealthy countries. In 2009 and 2010, ~75% of global installation volume was on small buildings that pay the highest price for electricity to traditional electricity networks.[26]

2. **Disproportionate profit impact:** The profit impact of a lost watt-hour from a small residential or commercial customer is typically 3X or more (sometimes much more) the profit impact of a lost large industrial customer. One reason for disproportionate impact is that prices for smaller volume users are often 50% to 200% above prices for larger users.[27]

3. **Multiplier effect:** In Germany, an upper-middle class household putting solar power on its roof reduces volume from the centralized network by ~5 million watt-hours per year, equivalent to more than $1,000/year of end-customer revenue.[28] There are also wholesale market impacts that further multiply the effect.

4. **Multi-year effect:** Customers who put solar power on their roofs stay with solar power for 20-plus years.[29] While there is some slight decay in electricity output over time, evidence from many installed panels suggest that they continue to function at a high performance level for at least two decades, often longer.

5. **Hard to recapture the customer:** Once a solar power system is on the roof, the cost of generating electricity is virtually zero for the customer (i.e., nearly no operating cost). As a result, it is nearly impossible for the centralized electricity system to recapture the volume.

6. **Very large scale:** There are already more than 1 million rooftop solar installations per year. In Germany, for example, there were more than 200,000 houses and small buildings that added solar power on roofs during the course of 2010. In aggregate, these solar power systems represent the largest addition of electricity capacity in Europe and North America, far larger than electricity capacity additions for coal, natural gas, nuclear, hydro or wind, or other technologies.[30]

7. **Declining traditional network volume:** The rapid rise of solar power has the potential to turn electricity markets with flat-to-low volume growth (e.g., much of Europe and North American) into negative volume trends. For example, Germany's total traditional electricity generation volume is down roughly 2% from 2007 to 2010 and mid-day generation volume is down roughly 10% on sunny days, with further declines likely in the future as more solar power is added.[31]

8. **Significant economic challenge for traditional network:** Solar power creates economic challenges for existing electricity networks by reducing the volume, decreasing the price and increasing the cost of the traditional network. In Germany, for example, small-scale rooftop solar power systems reduced revenue in the wholesale market by ~$2.5 billion and added generator costs of ~$2.5 billion. In addition, the end-customer incentives paid by utilities for generation of solar electricity equate to another ~$2.5 to ~$5.0 billion per year.[32] This delivers an economic hit of roughly $5 to $10 billion, significant in a market with roughly $100 billion in end-customer revenue.[33]

9. **Significant financial challenge for traditional network:** With less reliable volume, revenue and profit, the traditional electricity network will find it difficult to continue refinancing, making it impossible to invest into technologies, strategies or business models that might combat the

new entrant. While there is no way to predict when the financial markets will awaken to these risks, there are good reasons to believe that the challenges coming in Germany's traditional electricity sector by the start of 2012 may be enough to drive a broad-based reassessment of risk involved with financing traditional electricity infrastructure. The result might be significantly higher interest rates, shorter tenors and/or less liquidity for traditional electricity companies and also for traditional electricity infrastructure.

10. **Bigger, faster, worse:** All of the above happens at a scale that is larger and at a pace that is faster than most outside observers anticipate, leading to impacts that are worse than nearly anyone expects. Basically, solar power might actually become a trillion-dollar industry in the time period described by the executive at Sharp Solar back in 2002. In contrast the potential upside for solar power, traditional electricity companies and infrastructure may be forced to admit to significant uncertainty in volume, price and cost. Recognition of these economic challenges may be a trigger for driving reevaluation by financiers and may occur very quickly, perhaps within the next few years. The result might be a trillion-dollar solar power sector, but also hundreds of billion dollars in stranded traditional electricity assets and widespread challenges for delivery of electricity to less wealthy regions.[34]

Is it certain that this downside case will occur? No. Is it plausible? Absolutely. This plausibility means that there is a degree of insanity if smart people ignore the situation. The challenge for CEOs is that this seems like an unsubstantiated view because there is *little* solar power installed in many places. For example, in Boston, where I studied and worked, it is rare to see a solar power system. As a result, it is easy to dismiss the prediction of a negative network. However, this is exactly how negative network effects occur. When negative network effects begin, they look small, expensive and linear. It takes a long time for a new technology to reach a cost structure and volume that is capable of disrupting the existing network infrastructure. It takes even longer to realize that the growth is exponential not linear.

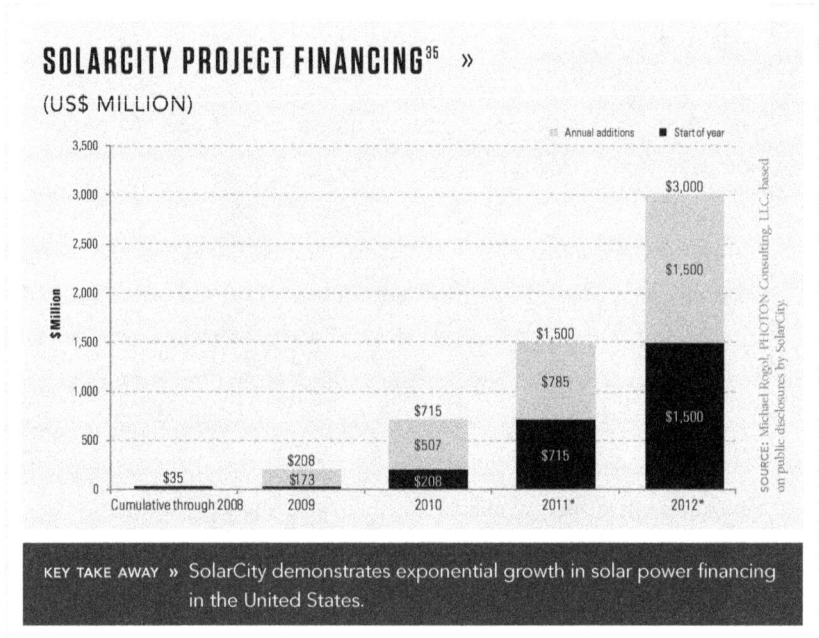

SOLARCITY PROJECT FINANCING[35] »
(US$ MILLION)

Legend: Annual additions ■ Start of year

	Cumulative through 2008	2009	2010	2011*	2012*
Start of year	—	$173	$208	$715	$1,500
Annual additions	$35	$208	$507 ($715)	$785 ($1,500)	$1,500 ($3,000)

SOURCE: Michael Rogol, PHOTON Consulting, LLC, based on public disclosures by SolarCity.

KEY TAKE AWAY » SolarCity demonstrates exponential growth in solar power financing in the United States.

Today, solar power is *already* at a cost and volume capable of significantly disrupting the traditional electricity sector in and around Boston. Today, solar power is *already* demonstrating that its growth is exponential not linear. The exponential rise of solar power in Boston has not occurred to date because other markets (most notably Spain, Germany, other European markets and even Ontario) have consumed much of the global supply of solar power panels at prices above what customers in the Boston area are willing to pay. But it *is* plausible that a major disruption (a genuine negative network effect) might occur in Boston within 2 to 5 years as solar power supplies become available at prices customers in this region are willing to pay. It *is* plausible that the monster might attack my hometown. The trajectory of companies in the U.S. such as SolarCity suggests this monster attack with exponential growth is coming quickly.

I hope a broad set of business leaders will realize what the rapid growth of solar power means: It is plausible that volume, price and cost of traditional electricity companies and infrastructure will come under attack from solar power much like the USPS is under attack from online bill pay. Further, my hope is that leading CEOs will aggressively seek paths to a heroic slaying or at least taming of the monstrous impacts of negative network effects both within electricity and in other areas in which negative network effects are rearing their heads. This hope is based on the strong conviction that solving these problems can be very profitable for companies and create value on a scale similar to Rockefeller and Carnegie. With this hope in mind, we turn to a more general description of negative network effects along with some short films to further highlight the prevalence of negative network effects in other areas of our infrastructure.

Chapter 3

AVOIDING EGO: INNER WORKINGS OF A WEAPON

Avoid having your ego so close to your position that
when your position falls, your ego goes with it.
—Colin Powell

AVOIDING EGO

"The Case of the Negative Network Effect" may have the ingredients to become a Sherlock Holmes economic murder mystery. In this detective thriller, there are claims that a newfangled weapon (the negative network effect) is attacking vulnerable infrastructure. Similar to asphyxiation, a negative network effect *may* lead to a painful death from declining volume. The mystery within this case comes from the fact that homicide has not yet occurred. It is a murder mystery *before* the murder. There are reports of attacks with a *potential* murder weapon. There is evidence the attacks cause injuries. The injuries are serious (e.g., dire fall in profit in U.S. mail infrastructure, similar evidence in European electricity infrastructure), but *may* not be life threatening. All suspects (online bill pay, rooftop solar power) are heroes with clean records, making it impossible to arrest with only circumstantial evidence of something that *might* happen in the future.

A murder mystery *before* the murder sounds like a great fit for any CEO with deep curiosity and relentless problem-solving skills. The "right" CEOs for this case will be very flexible thinkers with a lack of ego who are more interested in getting to the right answer than in being right. They will leverage others' ideas, adjust their own ideas and admit an initial answer is wrong if a better answer emerges. These characteristics of flexible thinking must be combined with precise observation and tight reasoning. For the CEO who possesses it, this is a combination reminiscent of Sherlock Holmes, one that is well suited to investigating the potential impacts of negative network effects before they have fully unfolded.

Solving the case hinges on the questions, "Do negative network effects exist?" and, "If so, how much of an impact will they make?" In other words, is the monster described in the preceding chapters real or imaginary and is it genuinely dangerous or just scary? Even at the outset of this case, it is unclear if definitive answers to these questions are possible.

POSITIVE NETWORK EFFECTS »
(SCHEMATIC OVERVIEW)

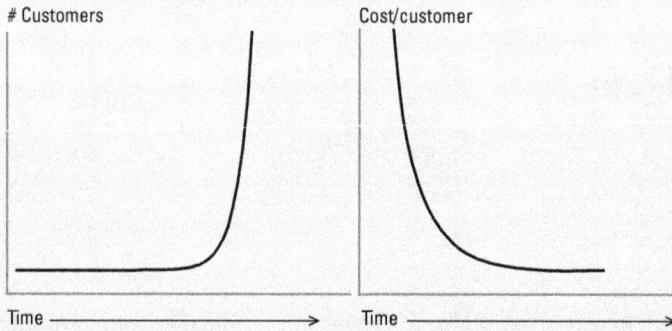

Customers

Cost/customer

Time →

Time →

SOURCE: Michael Rogol, PHOTON Consulting, LLC.
NOTE: All data are rough estimates.

KEY TAKE AWAY » Positive network effects are a familiar pattern.

What is clear is that "positive" network effects do exist and are important. Fundamentally, our energy, food, water and information flow through infrastructures that depend on positive network effects. In simple concept, a positive network effect occurs when an exponentially growing number of units run through a network with a high fixed cost structure. The "positive" result is that fixed cost per unit declines exponentially, making it less expensive per unit as the number of units grows. The declining cost enables suppliers to price at a level that attracts more customers. More customers mean more volume and revenue with only incremental cost, enabling more profit. The rising level of profit attracts low-cost capital to finance further expansion and operation of the network. This dynamic can be summed up as: Growth → lower cost → lower price → more customers → repeat.

Positive network effects are easy to understand in concept, and their impact is easy to believe in reality. One reason this is so easy to believe is that there are many examples around us of networks that create value. Historically, positive network effects enabled a broad set of infrastructure to achieve attractive economics, including Roman aqueducts, European railroads, U.S. interstate highways and the Internet, among many more. The massive value these networks generate often captures attention from the media and the public. In the last two centuries, the famous fortunes of Rockefeller (oil), Vanderbilt (railroads), Gates (software) and others were built by selling products that relied on positive network effects in a manner that created significant value for the suppliers while also delivering significant value to the end-user.

A second reason the impact of positive network effects is easy to believe is that a network's user can feel momentum as the network builds. Take the example of a customer using mobile telecommunications (i.e., a mobile phone). From the mid-1980s, a customer could easily observe a rising number of offers for mobile telecommunications products (service, phones, accessories), an increasing quality level of service (fewer dropped calls, stronger connection signal, more voice and data services) and a decreasing price. For example, the typical price of a 1-minute mobile phone call within the U.S. was $1.0 in 1980, $0.7 in 1990, $0.2 in 2000 and under $0.1 in 2010.[1] This equates to an 8% compound annual decline rate in the price of mobile telecom.

And price was just one aspect of "momentum" observed by the end-customer. Other aspects of momentum were even more tangible, such as smaller phones, which were enabled by technological advances, which were enabled by the rising profit of the sector, which were ultimately enabled by the rising volume of cell phone usage on a network with significant fixed cost. The overall point is that we all recognize the existence and impact of positive network effects because we all have experienced tangible benefits as volume on a network grows.[2]

While positive network effects appear to be real and impactful, this does not in any way prove that "negative" network effects exist. Even more, the way we *perceive* positive network effects may make it more difficult to believe in the possibility of negative network effects. This is because we *perceive* a close connection between end-user technology (e.g., car, phone) and the infrastructure network (e.g., road network, telecommunications network) on which that technology operates. This distinction between end-user technology and the infrastructure network is often blurred because technology transitions for

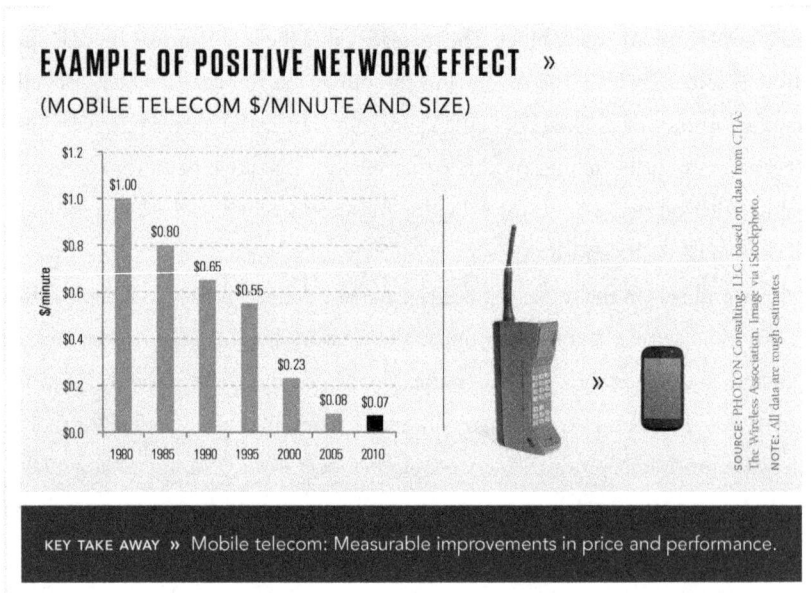

EXAMPLE OF POSITIVE NETWORK EFFECT »

(MOBILE TELECOM $/MINUTE AND SIZE)

SOURCE: PHOTON Consulting, LLC, based on data from CTIA-The Wireless Association. Image via iStockphoto.
NOTE: All data are rough estimates

KEY TAKE AWAY » Mobile telecom: Measurable improvements in price and performance.

end-users (when customers adopt a new technology that replaces an old technology) often occur at the same time that volume rises on the infrastructure network and at the same time that adjustments are made to increase the capacity, reach and resilience of the infrastructure network. These changes *seem* to move hand-in-hand.

Transportation provides an illustrative example. The major highways (e.g., Routes 1, 90, 93 and 95) where I grew up in the northeast U.S. are now crammed with cars on most days, but these were previously paths (e.g., Pequot Path) used by Native Americans before being widened, expanded and strengthened for use by horses, wagons and then cars.[3] The technology transitions (e.g., end-users adopting horses, wagons, then automobiles) in transportation took place over hundreds of years during which time rising volume (e.g., increasing number of passengers, vehicles, vehicle miles) necessitated modifications (e.g., widening, lengthening, strengthening) of the existing transportation network. During this period, end-user transportation technology changed at the same time that volume increased on the transportation network at the same time that the transportation network was significantly expanded.[4] In our minds, there is a perception that these are linked together.

Similarly, the adoption by end-users of new technologies in telecommunications (e.g., fax machines, mobile phones, Internet) in the 1980s and 1990s actually increased volume through existing telecommunications infrastructure networks at the same time that the telecommunications infrastructure networks were reaching a broader geography with more capacity and better functionality. Adoption of new telecommunications technologies by end-users brought more volume to the traditional telecommunications network (not just to new cell towers but also to existing land lines) and also drove modifications of the network (equivalent to widening, lengthening, strengthening). Again, there is a perception of linkage among end-user technology transition and rising network volume and infrastructure network modifications.[5]

At a fundamental level, we often assume a positive network effect because we have so often seen the pattern of end-user technology transitions (e.g., adopting a mobile phone) combined with volume growth for the infrastructure network (e.g., more minutes of use on the existing telecom network) combined with improvements made to the infrastructure network (e.g., broader geographic coverage with more capacity to handle the higher volume and better functionality to accommodate the more advanced mobile phone's capabilities). Yet this intimate familiarity with positive network effects that leads us to assume the positive network effect will occur also leaves us overlooking a core requirement: **Positive network effects are addicted to growth.**

Without volume growth on the existing infrastructure network, there is a problem with end-user technology transitions because network modifications are no longer economic. Here is the basic flow of the problem:

o   Without an exponentially increasing number of units on the infrastructure network, the fixed costs per unit do not decline.

- When costs do not decline for the infrastructure network, alternative technologies steal the highest price customers from an existing network.

- This is *not* alternative technologies stealing the highest price customers *within* an existing infrastructure network but *from* an existing infrastructure network so that high price volume leaves the existing infrastructure network.[6]

- When prices for the existing network infrastructure are raised to cover fixed costs over a smaller number of units, more customers leave and economic losses for the infrastructure network become clear.

- When economic losses for the infrastructure network become clear, capital markets refuse to provide low-cost financing to sustain the infrastructure network.

During the last decade, I have discussed this basic description of negative network effects with hundreds of executives, policy makers and academics. The general feedback from this broad group has been that this description of a negative network effect appears reasonable. However, when pressed, this broad group has been unable to come up with a single example of a negative network effect occurring in reality. They offer many examples of technology transitions and failed companies, but no examples of genuine negative network effects. I am still searching for real-world examples and am asking for help from others. In a way, the CEOs reading this book may think that I am asking them to be afraid of the boogeyman.

Despite the lack of concrete case studies, the general consensus is that it is *plausible* for a negative network effect to occur when an end-user technology transition causes an exponential decrease in the number of units running through a traditional infrastructure network's fixed cost structure. The "negative" result is that fixed cost per unit increases exponentially, making it more expensive per unit as the number of units decline. Here is a schematic figure to demonstrate the point.

NEGATIVE NETWORK EFFECTS »
(SCHEMATIC OVERVIEW)

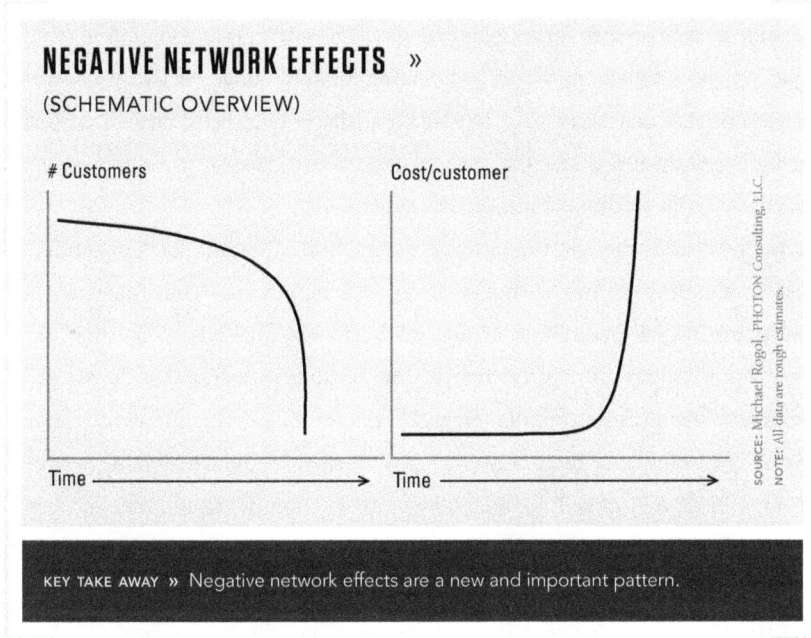

Customers

Time ⟶

Cost/customer

Time ⟶

SOURCE: Michael Rogol, PHOTON Consulting, LLC.
NOTE: All data are rough estimates.

KEY TAKE AWAY » Negative network effects are a new and important pattern.

In contrast with positive network effects, the importance of negative network effects may be easy to understand in concept, but difficult to believe in reality because there are no obvious examples around us of infrastructure networks that actually have a decline in volume. Yes, we observe many examples of old technologies being discarded by end-users for new technologies. Yes, we observe many companies going out of business because they are unable to adjust. Yes, we observe many failures in parts of our infrastructure (e.g., breach of levees in New Orleans in 2005, collapse of the Mississippi River Bridge in Minnesota in 2007). However, no, we have not observed the fundamental failure of a broad infrastructure network. This is what makes the examples of the U.S. mail infrastructure network and European electricity infrastructure networks so important.

The preceding chapters highlight two fundamental failures of broad infrastructure networks that are works in progress. Chapter 1 shows how rapid adoption of online bill pay is leading to exponentially higher fixed cost per unit for the traditional mail delivery network. Chapter 2 shows how the rapid adoption of rooftop solar power is following the same dynamic for traditional electricity networks. These observations point to a fundamental disconnect between end-customer technology transitions and the growth necessary to sustain a positive network effect in specific traditional infrastructure networks.

In both cases, though, the ultimate destruction of traditional companies and traditional infrastructure has not occurred, at least not yet. While there are signs that point in that direction, they are only signs not final results. Further, there are many examples of negative impacts on existing networks that do not go all the way to penultimate capitulation and death for the existing infrastructure network. One example is the continuing use of existing landline telephone networks despite the advent of cellular telephony, introduction of the Internet and bankruptcy of companies like AT&T. These examples of major technology transitions without death for the underlying infrastructure network suggest that technology transitions are often much less harrowing than the horror movie analogies described previously.

This is an unsatisfying situation despite the potential for a thrilling plot. We have a potential murder-in-the-making, but are unable to figure out if the murder will occur or if something else will happen to prevent it. Given the potential magnitude of the implications if infrastructure networks fail, I hope to convince CEOs and other executives to engage this topic with the mindset of Sherlock Holmes. I am asking them to do this after nearly a decade researching this topic myself. Based on this research, here is where I come clean on my answers to the questions presented earlier in this chapter.

"Do negative network effects exist?"

My best guess is, yes, they do exist. Like an astronomer predicting a comet's path, my expectation is that we will be able to observe negative network effects with the naked eye by 2012–2013 when negative volume growth in several infrastructure networks becomes a recognized pattern instead of a perceived one-time problem. While no one can accurately predict the future, my sense is that the volume-related problems facing U.S. mail infrastructure, European electricity infrastructure and possibly other infrastructure networks will become apparent and irrefutable by 2014, possibly before the start of 2013 and plausibly even before the start of 2012. This is not a new prediction. I have been studying this since the early 2000s and writing this since 2006. What's new is that I am asking others with more experience to join the case. I am asking others to join the case based on my preliminary answer to the second question.

"How much of an impact will negative network effects make?"

My best guess is that negative network effects *may* fundamentally undermine energy, food, water and communications infrastructure and companies in developed economies in the coming years. It is *plausible* that this will occur.[7] It seems reasonable that the adoption of new energy, food, water and communications technologies at or near the point-of-use will reduce usage of existing energy, food, water and communications infrastructure. This is not just reducing the growth rate (e.g., going from positive 10% annually to positive 5% annually) but actually reducing the volume going through these infrastructure networks (e.g., going from 10 billion watt-hours to 9.5 billion watt-hours, equal to *negative 5%*). If this occurs, it then seems reasonable that the companies operating these networks and the infrastructure networks themselves will face significant economic and financial challenges.

By the end of 2010, electricity generation in Germany's traditional electricity infrastructure (in this case, Germany's nonsolar power electricity infrastructure) was down approximately 25% on weekdays that were sunny and 34% on weekends that were sunny. In 2011, this share of solar power generation (i.e., the reduction in traditional electricity generation) is rising quickly, to approximately 34% of weekday peak and 46% of weekend peak during sunny hours.[8] This is a major problem for the traditional electricity infrastructure, one that is getting worse as more solar power is added each month.

SOLAR POWER SHARE OF PEAK ELECTRICITY GENERATION IN GERMANY, 2010 VS. 2011[9] » (%, GERMANY)

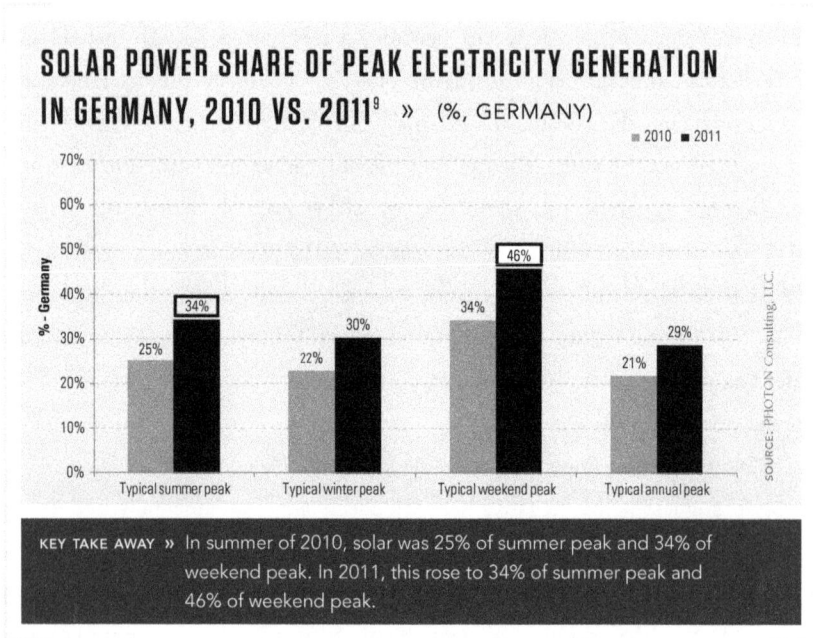

KEY TAKE AWAY » In summer of 2010, solar was 25% of summer peak and 34% of weekend peak. In 2011, this rose to 34% of summer peak and 46% of weekend peak.

This is not about people purchasing cell phones at a time when nearly everyone was expanding the volume of phone usage (number of minutes per month per user), when more people were purchasing multiple homes with multiple landline phones (number of private phones per person) and when businesses were also adding landline phones (number of business phones per person). This is about the *plausibility* of total electricity volume within the existing electricity network declining year-on-year for many years to come. **This is about the plausibility of the same pattern occurring at about the same time for other traditional energy, food, water and communications infrastructure throughout a large portion of the wealthy world.**

In the case of mobile telecom in the U.S., the traditional phone companies and phone networks faced challenges, but the challenges were balanced by positive trends in volume. In the case of electricity (and possibly other areas of traditional energy, food, water and communications), there is a real case to be made that many zip codes throughout the developed world will see declining consumption from traditional electricity infrastructure networks for many years to come due to flat population, flat economies, rising energy efficiency and rising use of on-site generation (i.e., distributed generation) such as rooftop solar power. **If this occurs, the results would be catastrophic for the economics and financing of the existing electricity infrastructure. If this occurs, it raises a specter of similar impacts for much of our traditional energy, food, water and communications infrastructure.**

The reasons I am concerned about more than just the mail infrastructure network and the electricity infrastructure network is that other traditional infrastructure networks show anecdotal signs of negative network effects. For example, similar dynamics can also be observed within our food infrastructure. As previous chapters pointed out, negative network effects are a targeted attack on the best customers. In the food sector, the best customers are those that purchase high-price, high-profit food. These best customers within our food system are taking actions that suggest negative network may be under way. Take a look at produce, which is the most profitable section of many high-end grocery stores (hence located near the front door when shoppers arrive). Within the produce section, mushrooms are among the highest-price and highest-profit foods on a dollar per pound basis.

There are now examples of point-of-consumption production technologies for high-price products (e.g., mushrooms) that enable production to move closer to the point of consumption (e.g., near your kitchen and dining room). What's interesting is that the customer adoption patterns of these point-of-consumption production technologies are similar to the rapid adoption of rooftop solar power that is installed close to the point of consumption for electricity (e.g., near your air-conditioner and TV). Patterns in other industries suggest that there would be a rapid take-off in end-customer adoption of technologies that enable on-site production of high-price produce (e.g., mushrooms) at a cost to the end-customer that is below the grocery store price.

I have been discussing this with my wife, Susan, who is a chef and also someone who knows a lot about household food preparation. Susan went to culinary

AVERAGE RETAIL PRICE FOR FRESH VEGETABLES »
(\$/POUND)

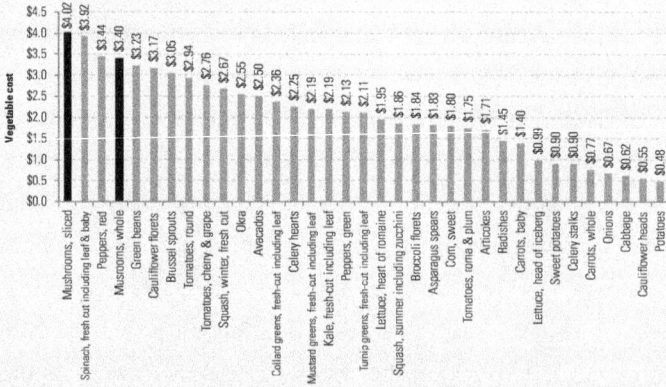

Vegetable	Cost
Mushrooms, sliced	$4.02
Spinach, fresh cut including leaf & baby	$3.92
Peppers, red	$3.44
Mushrooms, whole	$3.40
Green beans	$3.23
Cauliflower florets	$3.17
Brussel sprouts	$3.05
Tomatoes, round	$2.94
Tomatoes, cherry & grape	$2.76
Squash, winter, fresh cut	$2.67
Okra	$2.55
Avocados	$2.50
Collard greens, fresh-cut including leaf	$2.36
Celery hearts	$2.25
Mustard greens, fresh-cut including leaf	$2.19
Kale, fresh-cut including leaf	$2.19
Peppers, green	$2.13
Turnip greens, fresh-cut including leaf	$2.11
Lettuce, heart of romaine	$1.95
Squash, summer including zucchini	$1.86
Broccoli florets	$1.84
Asparagus spears	$1.83
Corn, sweet	$1.80
Tomatoes, roma & plum	$1.75
Artichokes	$1.71
Radishes	$1.45
Carrots, baby	$1.40
Lettuce, head of iceberg	$0.99
Sweet potatoes	$0.90
Celery stalks	$0.90
Carrots, whole	$0.77
Onions	$0.67
Cabbage	$0.62
Cauliflower heads	$0.55
Potatoes	$0.48

SOURCE: U.S. Department of Agriculture. "How Much Do Fruits and Vegetables Cost?", 2011. www.ers.usda.gov/Publications/EIB71/EIB71.pdf.

NOTE: All data are rough estimates.

KEY TAKE AWAY » Mushrooms are among the most expensive produce in grocery stores.

school and worked in and then ran restaurants. She was recruited to move to New York to work directly for Martha Stewart in the early 2000s. She was on the team that launched *Everyday Food*. It's a small format magazine that went to 1 million subscribers within a year. This is a big number for the magazine industry and her team received the award for best magazine launch of 2003. The concept of the magazine was to make the process of preparing food easier (a concept of high relevance to this discussion). The magazine was a hit because it centered on user-friendly food that could be found in nearly all grocery stores, carried home in one shopping bag and prepared in a short amount of time. The magazine was useful and popular. Eventually other magazines copied this formula, but *Everyday Food* still does well.

Susan no longer works with the magazine (she moved to Boston 5 years ago to pursue an MBA), but she enjoys staying on top of food trends. In an article from the *Wall Street Journal*, Susan learned that Whole Foods has had success selling grow-at-home food kits. She sent me the article as an example of a trend she is observing:

Whole Foods began selling the kits out of a single store in Berkeley last spring. The kits sold out fast . . . Now, they are sold in more than 150 Whole Foods stores. "At Christmas we put it out as a potential gift item and we couldn't get enough and sold out . . ."[10]

What's interesting to me is the pattern of adoption. This sounds so familiar. The wording is very similar to my interviews with companies in Japan during 2002 to 2004 discussing the adoption of rooftop solar power. In Japan at that time, executives regularly told me that that they were surprised by rooftop solar power. They would typically say, "It is performing much better than we expected." This is exactly what Whole Foods is saying about grow-your-own mushrooms.

My take is that "distributed" food production (i.e., food produced at the point of consumption) seems to have many similar patterns as the distributed electricity production and distributed water purification patterns I have been studying since starting my never-to-be-finished PhD 9 years ago. The basic pattern is compound exponential growth when the internal rate of return (IRR) of the all-in cost of the distributed infrastructure goes above the end-customer's discount rate. In nonfinancial commonsense terminology, this basically means that people buy when they get a good deal. This normally happens for higher-priced customers/ products first. In the early 1990s, Japan's residential electricity was the first market tipping for rooftop solar power because they had very high electricity price combined with low interest rates. The result was very fast uptake of rooftop solar power once the cost of solar power to the end-customer was below the price of network electricity.[11]

Given this pattern, it's not surprising that the consumers at higher-end grocery stores (and especially the higher-end products within those stores) in the U.S. would start to show a similar pattern of "tipping." I have heard food experts talk about very fast growth for do-it-at-home mushrooms, tomatoes, honey and other high-price grocery story products. There are many anecdotes of high-price foods that normally flow through traditional infrastructure network (big farm → big distribution center → big grocery store) being displaced by a customer technology transition to at-home production for some small but high-price products of their food supply. (For another example, see the data on farmers markets in Chapter 5.)

And it is not just the high-end products of high-end grocery stores that are being displaced. Food trucks in LA, New York and many other U.S. cities have moved beyond traditional hotdogs and sandwiches into higher-end cuisine. The quality (not the price, but the quality) of these food trucks is what captures my attention. Historically, mobile food trucks were able to get much closer to more consumers than nonmobile restaurants, but the food trucks were not considered substitutes for high-end restaurant cuisine. This has changed. Food trucks have drawn significant attention from food critics, who admire the food, and ire from traditional restaurant owners, who watch their customers being taken by the rolling competitors. The restaurant review leader, Zagat, has even added coverage of food trucks to its traditional guidebooks and begun beta testing a separate food truck review service. Similarly, food truck chef Roy Choi won the 2010 "Best New Chef" award from *Food & Wine* magazine.[12]

The combination of food anecdotes I have collected over the last few years suggests that distributed food (produced closer to the end-customer's point-of-consumption) may be making inroads into the volume (especially the highest price, highest profit volume) of our centralized food infrastructure. Substituting a $6 gourmet food truck meal for a $36 high-end restaurant meal may make good economic sense for the end-customer, but it is painful for the existing infrastructure (in this case, the firmly established restaurant and its established supply chain). At small scale, this is just a small example of competition, but at larger scale this has the potential to fundamentally undercut the existing food-service infrastructure.

Please do not misunderstand the logic of a bunch of homegrown mushrooms taking down the basic underpinning the U.S. food infrastructure network. Similarly, please don't misunderstand the logic of food trucks being part of the same trend. The homegrown mushrooms, tomatoes, honey, eggs, chicken and other produced-near-the-point-of-consumption foods will *not* replace a significant portion of the U.S. food supply or U.S. food consumption anytime soon. Similarly, food trucks will *not* replace a significant portion of restaurant-supplied meals anytime soon.

However, it *is* plausible that produced-near-the-point-of-consumption foods will cause significant harm to the traditional infrastructure network even without displacing a significant portion of the volume. These produced-near-the-point-of-consumption foods only need to displace enough volume going through the traditional food infrastructure so that spending on food from the traditional food infrastructure in specific zip codes shifts from slightly positive growth to slightly negative growth. In other words, a small shift in volume away from the traditional infrastructure network may actually mean the difference between positive growth and negative growth.

For many towns in the wealthy world, this shift from positive spending growth on the traditional food infrastructure network to negative spending growth on the traditional food infrastructure network might require displacing only 5% of food volume supplied to end-customers in a small geographic region. Much like the USPS infrastructure relying on First-Class letters for the bulk of its profit, the centralized food distribution networks rely on high-end food products purchased by high-end customers for a disproportionate share of profit. Rooftop solar power was approximately 5% of total electricity consumption in Germany in 2010, but was already creating havoc on the fundamentals of Germany's traditional electricity sector.[13]

As mentioned earlier, growth in volume is very important for a traditional infrastructure networks. The importance of growth ultimately comes down to a single cell within a very large spreadsheet. Investors (equity) and lenders (debt) use detailed net present value (NPV) calculations to assess potential investments and loans to maintain and expand infrastructure networks. Within a large spreadsheet model to calculate NPV, there is a single cell to estimate the "residual value" of the asset at the end of the modeling period. This "residual value" is basically an estimate of the value of the asset in the long-term future, typically after 10 years.

Within this single cell is a formula that includes an expectation for growth in cash flow after year 10.

If the expectation is that cash flow will have positive growth after year 10, the NPV calculation is typically much higher than if the expectation is that cash flow will be flat or falling after year 10. The change of this number within one formula within one cell can wipe out a significant portion (often half) of the net present value of the infrastructure asset. Basically, without growth, the asset is worth much, much less in the eyes of people financing the asset. Anything that might turn positive volume growth to negative volume growth would significantly reduce expectations for future cash flow growth, which would significantly reduce the NPV of the asset, which would make assets like this much more difficult (i.e., expensive) to finance.

Without growing volume, it is harder to drive down cost (i.e., you are not able to spread fixed cost over a larger base), which requires raising prices to maintain profit in the face of even modest inflation in feedstock or maintenance costs. The higher prices make it easier for distributed infrastructure to capture more volume from the traditional centralized network. This in turn makes it even more difficult (i.e., expensive) to finance the traditional centralized network. This is a *potential* death spiral that is only now becoming apparent.

What normally happens when there is a pattern of "distributed infrastructure" replacing something from "centralized infrastructure" is that I hear lots of excuses for why the pattern is an anomaly or why the pattern can be ignored. And adoption of distributed infrastructure normally happens in a way that is easy to ignore because it is too small or too expensive. That is until you look more carefully at the growth rate (very fast) and do more detailed accounting/analysis of the end-customer economics. In the case of grow-it-yourself food, I often hear executives explain it away by saying, "It's niche" or "It's only due to the slow economy." I don't think so. I think it's because the all-in cost of some types of grow-it-yourself distributed food has fallen below the all-in price of centralized food (big farm → big distribution chain → big grocery store). When the economics become good enough, people adopt. And they adopt with very fast growth rates that appear similar to the rooftop solar power example. And professional supply chains grow quickly to meet the surging demand.

In the case of the mushrooms, the *Wall Street Journal* reported, "A variety of mushrooms that were once hard to find now can be grown at home. It's more convenient—and more entertaining—than buying mushrooms at the store. But it doesn't necessarily save money."[14] I strongly suspect that a more detailed analysis of the economics would show that it does save money if you compare the all-in cost for homegrown versus the all-in price at the store. This is only a guess, but it is based on doing analysis on similar patterns in energy and water purification. My hope is to conduct more detailed analysis soon while also convincing sharp thinkers to make their own assessments.

What really catches my attention (and I think should catch the attention of CEOs across industries) is that magazines are now thinking about this pattern. Their advertising teams are looking to sell ad space to companies in the distributed food supply chain and initial signals say that they are finding companies, often large companies, with whom they are now discussing advertising deals. The main reason that the magazine doesn't have these ads already is because "the distributed food companies are sold out and don't feel a lot of pressure to advertise." Sound familiar? Probably not to you, but this reminds me of rooftop solar power from 2004 through mid-2008.[15]

You might ask, "Why am I bugging CEOs in a broad set of industries with this?" There is a simple reason: I really want them (and others) to consider (really consider not just politely read), what the impact of negative network effects is for traditional centralized infrastructure. My gut says that a new era is being born, even if most people do not recognize it yet. This is the era of distributed infrastructure industries. I don't know if my gut is right, but I want to bring light to this topic so that these CEOs (and others) can let it bounce around in the back of their heads. This is important because *if* they see this as the dawn of a new era of distributed infrastructure instead of disconnected stories about online bill pay and rooftop solar power and self-grown-mushrooms, it may change the focus of their companies and their careers.

To be clear, I am not the only one who thinks this way. For example, the CEO of a global, multibillion-dollar company involved in the semiconductor industry wrote this to me recently:

> Distributed is the way to go in all industries and walks of life. The world moved to centralized manufacturing, markets, utilities, etc. because that was the only way human beings knew how to cut costs (through scale and scope). Societies before the industrial revolution all depended on "distributed" means and most of us want to go back there. We have a deep psychological need to be independent, to control our destiny. The blocking mechanism is cost. Once that barrier is overcome massive movement to distributed structures will ensue.[16]

Please note: This is the CEO of a global semiconductor company, not a crunchy, organic green farmer. The "massive movement" he mentions is already starting in areas where costs for distributed infrastructure are below the price of high-end products available from traditional centralized infrastructure networks. This is what we observed in First-Class mail converting to online bill pay and high-price electricity customers converting to rooftop solar power. **The battle is between traditional centralized infrastructure (built on a foundation of centralized low-cost manufacturing leveraging economies of scale combined with low-cost distribution leveraging positive network effects that overcome inefficiencies in not perfectly matching production with use) and new distributed infrastructure (being built on a foundation of customized-to-use production with a long-term commitment at the point-of-use).**

This movement creates potentially huge business opportunities. This is particularly true if the adoption pattern is similar *and* predictable across distributed infrastructure technologies. **Predictability of the adoption pattern is the key.** If you know (1) at what price customers turn on and (2) at what rate they grow once turned, then you have a much more predictable business that will justify more significant up-front investments. These moneymaking business opportunities and what they mean for CEOs across industries will be a focus through the remainder of the book.

Before turning to these opportunities, though, it is important to turn back to this mystery of a murder-before-the-murder. This chapter and the preceding chapters have a dark tone, emphasizing attacks, monsters and murders. It is important to understand these dark tones even while looking for bright opportunities. The dark side of negative network effects is that those who are economically less well-off are more vulnerable to the transition. Members of society who are wealthy or whose neighbors are wealthy are more likely to smoothly transition from centralized infrastructure to distributed infrastructure. Members of society who are poor and whose neighbors are poor are more likely to face a rocky transition. The *Wall Street Journal* and many other media outlets have reported on issues facing poorer rural communities. Their coverage focuses on the USPS playing an important role by linking remote areas with the rest of the country while also having significant loses. With loses growing, the USPS is seeking to close roughly 2,000 locations in addition to the roughly 500 that it wanted to close from the end of 2010. This is bad news for rural areas that rely on the post office for community, identity, commerce and deliveries. In particular, many elderly in rural areas rely on the USPS for pharmaceutical deliveries. In some locations, "The local school closed years ago and reliable cable, Internet and cell phone reception has yet to arrive."[17]

Basically, the transition from traditional centralized infrastructure to new distributed infrastructure has potential to create a lot of pain for a lot of people. My biggest worry is that we may misdiagnose a *real* negative network effect as a common technology transition. Confusing a negative network effect with a common technology transition would be like misdiagnosing a broken toe instead of an epidemic disease. Misjudging a negative network effect as a common technology transition would be like arresting someone as an isolated pickpocket instead of as a crime boss whose organization threatens the safety of all citizens. It is important to recognize the potential downside impacts from potential transformation to distributed infrastructure. I am dedicated to helping minimize these downsides. My hope is that others will aggressively pursue the positive sides of negative network effects by building distributed infrastructure businesses and industries while also addressing the downsides.

"The Case of the Negative Network Effect," is still open. There is reasonable evidence to suggest negative network effects may be a root cause underpinning significant challenges facing our society, but this is just a hypothesis. Additional proof is necessary, but action is required before all necessary data is available. This is where executive skills (including an ability to make a core decision with limited information) combine with flexible thinking (including an ability to revisit a core decision as more information becomes available). It seems that the path to capture the positive sides of negative network effects will require aggressively pursuing action but also relentlessly questioning if and how to improve.

Per Colin Powell's advice, this is a situation in which it is important avoid having my position too close to my ego. If new evidence emerges to suggest that negative network effects are not real or that they are not impactful, I look forward to changing my position.[18] In the meantime, I will be aggressively pursuing the positive sides of distributed infrastructure while attempting to minimize the negatives. My hope is that other CEOs will do the same.

Chapter 4

POSITIVE SIDES OF NEGATIVE NETWORK EFFECTS

I will tell you how to become rich. Close the doors. Be fearful when others are greedy. Be greedy when others are fearful.
—Warren Buffett

POSITIVE SIDES OF NEGATIVE NETWORK EFFECTS

I am fearful about negative network effects, but I would not bother a group of CEOs with a story containing only depressing downsides. No way. CEOs need an upbeat adventure film. Tarzan in the business jungle. India Jones finding lost treasure. Rocky fighting to become champion. The narrative is, "Overcoming monumental obstacles to achieve greatness." CEOs will triumph over nearly any impediment with relentless determination, but only if they believe they are on the way to a worthy prize. When I approach senior executives about negative network effects it is because of the massive positives involved with the adventure. If they can answer the question, "Where is value?" then their efforts to capture it will be spirited.

This is my task: To convince CEOs that there is significant value from negative network effects. Over the last couple years, I have had an ongoing dialogue with numerous CEOs focusing on this topic. At first, many were skeptical. In response, I laid out snapshots of "things that don't make sense" in a way that caught the curiosity of the 10-year-old in them. As Ronald Reagan once said, "One picture is worth 1,000 denials." Over time, they could not ignore the pictures I showed them from the professional world in which I live. These "pictures" show the tantalizing, positive sides of negative network effects.

Let's start with an enticing picture that is worthy of a CEO's attention: A massive profit pool. CEO's like big profit pools more than big-stakes gamblers like the Rehab pool at Hard Rock Hotel in Las Vegas. Solar power is a sexy party pool for ambitious CEOs, so let's dive in.

PARTY POOL »
(ILLUSTRATIVE)

SOURCE: iStockphoto.

KEY TAKE AWAY » Big profit pools on the positive sides of negative network effects.

In the solar power sector, the companies that produce the feedstock for solar power (high purity silicon) capture a significant portion of the value. I figured this out in 2003 during my never-completed PhD research and then spent a lot of time getting to understand this goldmine within the solar sector. To do this, I began visiting many silicon plants, meeting with a broad set of executives and organizing silicon conferences at which all the large producers presented on a public stage before striking business deals in private rooms. In the years since, I have come to know them fairly well.

Perhaps the most striking aspect of silicon producers is that they are among the most profitable group of companies in the world. This group includes eight companies that account for more than 80% of total global supply. These companies are Hemlock (U.S. company owned largely by Dow Corning), Wacker (German company publicly listed in Germany), OCI (Korean company publicly listed in Korea), GCL (Chinese company publicly listed in Hong Kong), REC (Norwegian company publicly listed in Norway), MEMC (U.S. company publicly listed in U.S.), Tokuyama

(Japanese company publicly listed in Japan) and LDK (Chinese company publicly listed in U.S.).[1] What's interesting about these companies is that most are publicly listed, resulting in significant transparency in their volumes, prices, costs and profits. In other words, an outsider can get insight into these companies without any confidential information.

Take Wacker as an example. Among many details disclosed each quarter, Wacker reports revenue and profit for its Polysilicon Division.[2] Since 2005, the Polysilicon Division increased revenue from 288 million euro in 2005 to 1.4 billion euro in 2010. At the same time, its operating profit (Earnings Before Interest and Taxes, EBIT) grew from 66 million euro to 580 million euro. This is a 5X increase in revenue and an 8X expansion of profit in 5 years, with an operating profit margin in 2010 of 42% compared to 23% in 2005. This is a remarkable trend, with operating profit (EBIT) margin above 21% *every* quarter from 2005 through 2010, even during economic recession. And the trend looks likely to continue in 2011, with analysts estimating Wacker's silicon operations will generate a 43% operating profit (Earnings Before Taxes, EBT) margin in 2011.[3]

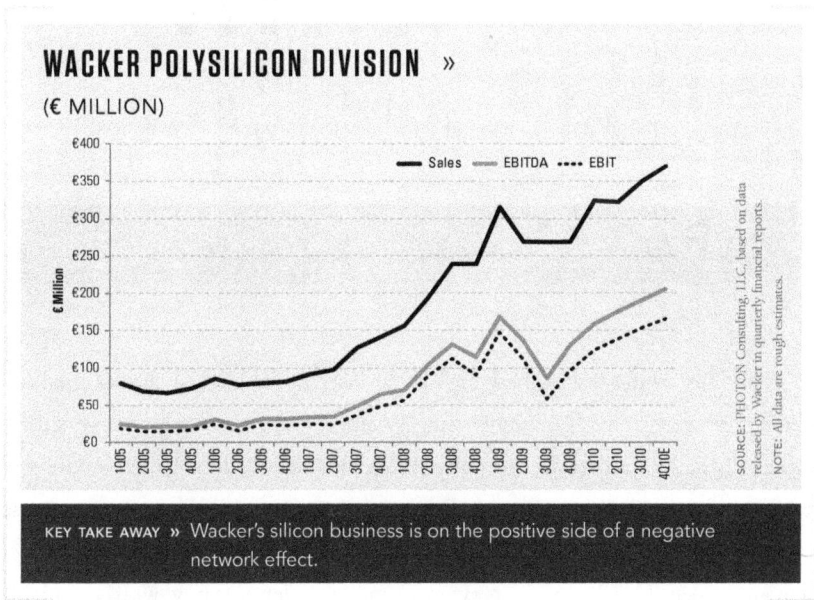

WACKER POLYSILICON DIVISION »
(€ MILLION)

SOURCE: PHOTON Consulting, LLC, based on data released by Wacker in quarterly financial reports.
NOTE: All data are rough estimates.

KEY TAKE AWAY » Wacker's silicon business is on the positive side of a negative network effect.

71

Similarly, the analysts estimate high 2011 operating profit (EBT) margins from silicon activities (not the overall companies but just polysilicon) of Hemlock (50%), OCI (41%), GCL (43%), REC (41%), MEMC (41%), Tokuyama (24%) and LDK (31%). These are remarkably high profit margins. To put this in perspective, operating profit margins for the world's leading companies in the most recent year are 6% for Walmart, 9% for GE, 14% for Exxon Mobil, 33% for Goldman Sachs and 39% for Microsoft.[4] The next figure is a stunning example of just how profitable silicon companies are. Basically, this group of solar companies looks remarkably profitable even in comparison to a group of the world's best companies.

OPERATING PROFIT MARGIN—SI VS. WORLD LEADERS »

(%)

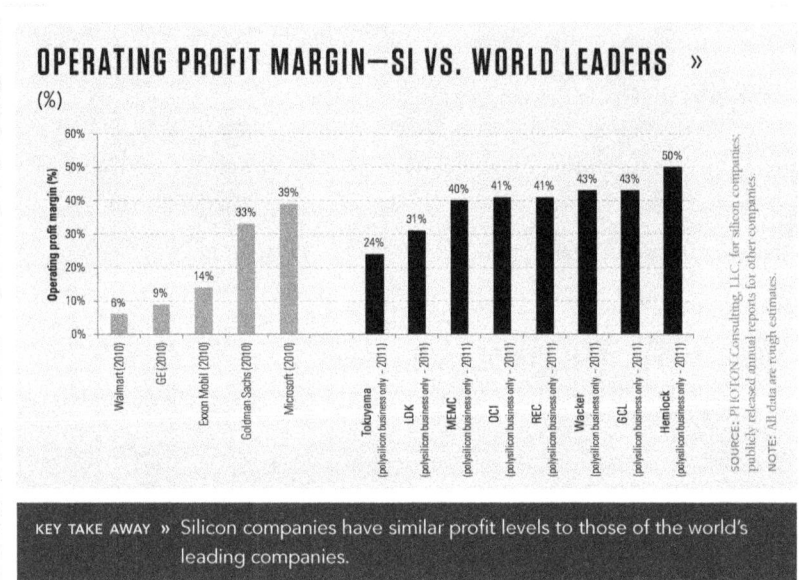

KEY TAKE AWAY » Silicon companies have similar profit levels to those of the world's leading companies.

Economists will say that these high profit margins are likely to attract attention and competition. This is certainly true. Whereas there were only 6 main silicon producers in 2005, there are currently 199 companies producing or attempting to produce silicon. That's a pretty significant jump. But it's more than just an increase in the number of companies. It is also easier to produce silicon. Whereas production equipment for silicon was previously special order, it is now widely available from equipment vendors who offer to build an entire silicon production facility if you will pay them a reasonable price for the facility. The barriers to entry have decreased significantly.

Basically, market forces appear to be working because the high profit margins of this sector are attracting new entrants and significant growth. Specifically, the total volume of silicon production has grown from 32 thousand tons in 2005 to 164 thousand tons in 2010 and likely more than 220 thousand tons in 2011. During the same period, silicon supply used in the solar sector increased from 13 thousand tons in 2005 to 195 thousand tons in 2011, equal to a 57% compound annual growth rate. In other words, the high profit margins of silicon have clearly attracted attention and additional supply.[5]

GLOBAL HIGH-PURITY SILICON PRODUCTION »
(THOUSAND TONS/YEAR)

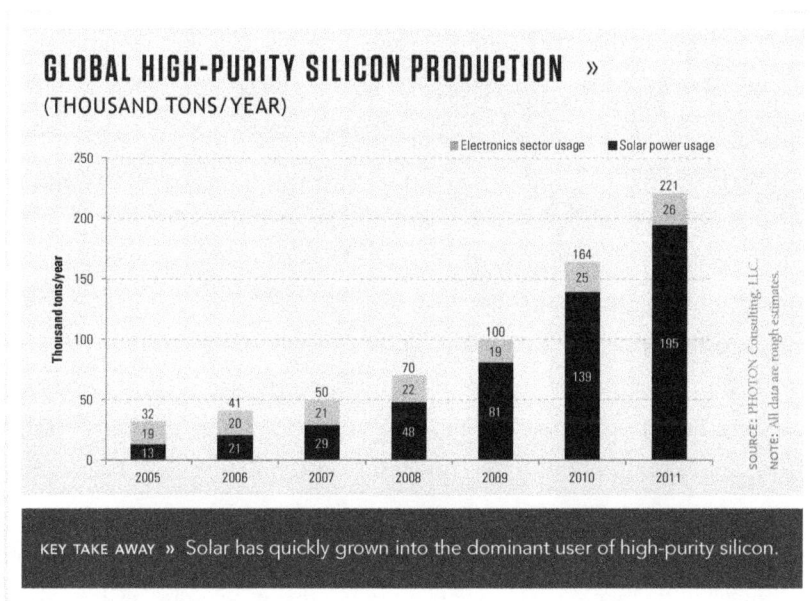

KEY TAKE AWAY » Solar has quickly grown into the dominant user of high-purity silicon.

Isn't it hard to believe that silicon companies have remained highly profitable despite (1) lower barriers to entry, (2) significant new entrants and (3) rapid production growth? Don't economists normally show that more supply leads to more competition with lower prices and lower profit margins? This does not make sense. A closer look shows that prices are far above costs, which is why production continues to rapidly expand. More specifically, outside-in estimates (i.e., done by analysts without any confidential information) for the all-in cost of manufacturing polysilicon for the eight largest producers ranges from roughly $25/kg to $45/kg, depending on the producer. This compares to average selling prices for producers of $55/kg in 2010 and $55/kg again in 2011. Overall, this equates to an average 33% operating profit margin in 2011 for all silicon producers.[6] It is hard to believe that such high profit continues in the face of such rapid supply expansion.

Reasonable people might suspect that silicon producers are an anomaly, that they are a unique group within the solar power sector. While silicon producers are certainly a special group, there are also many examples of other solar power companies that are highly profitable. For example, in the most recent quarter, there are at least 15 publicly listed solar power companies with revenue of $200+ million per quarter that do not derive a significant portion of their business from producing silicon. These companies include: Canadian Solar (Chinese company listed in U.S.), First Solar (U.S. company listed in U.S.), Gintech (Taiwanese company listed in Taiwan), JA Solar (Chinese company listed in U.S.), Jinko Solar (Chinese company listed in U.S.), Q-Cells (German company listed in Germany), Renesola (Chinese company listed in U.S.), SAS (Taiwanese company listed in Taiwan), SMA (German company listed in Germany), Solarfun (Chinese company listed in U.S.), SolarWorld (German company listed in Germany), SunPower (U.S. company listed in U.S.), Suntech (Chinese company listed in U.S.), Trina (Chinese company listed in U.S.) and Yingli (Chinese company listed in U.S.). As with the silicon producers, these solar companies are publicly listed with significant transparency that enables an outsider to get insight into their volumes, prices, costs and profits without any confidential information.

Take a close look at this table. This is nearly $8 billion in quarterly revenue with nearly $1.5 billion in quarterly operating profit. A 19% operating profit margin makes this group of companies less profitable than the silicon makers'

REVENUE AND OPERATING PROFIT FOR SELECT SOLAR POWER COMPANIES » (OVERVIEW)

	Revenue ($ million/quarter)	Operating profit ($ million/quarter)	Operating profit margin (%)	Reporting period (Timing)
Canadian Solar	$377	$40	11%	3Q10
First Solar	$798	$212	27%	3Q10
Gintech	$248	$45	18%	3Q10
JA Solar	$589	$89	15%	4Q10
Jinko Solar	$215	$56	26%	3Q10
Q-Cells	$527	$38	7%	4Q10
Renesola	$359	$86	24%	3Q10
SAS	$200	$47	23%	3Q10
SMA	$808	$255	32%	3Q10
Solarfun	$327	$59	18%	3Q10
SolarWorld	$462	$70	15%	3Q10
SunPower	$937	$135	14%	4Q10
Suntech	$744	$63	8%	3Q10
Trina	$642	$145	23%	4Q10
Yingli	$610	$141	23%	4Q10
TOTAL - QUARTERLY	$7,843	$1,481	19%	3Q10 or 4Q10
TOTAL - IMPLIED ANNUAL	$31,372	$5,924	19%	12 MONTH ESTIMATE
AVERAGE - PER COMPANY PER YEAR	$2,091	$395	19%	12 MONTH ESTIMATE

SOURCE: PHOTON Consulting's data platform, The Wall, and FactSet. NOTE: All data are rough estimates.

KEY TAKE AWAY » Sizeable revenue and profit pools for solar power companies.

33% operating profit margin, but still highly profitable when compared against nearly any other group of companies in the world. In other words, solar power companies without significant silicon production capacity are an additional deep profit pool. Even without any growth, simply multiplying these quarterly results by 4 (i.e., 4 quarters per year) equates to more than $31 billion in annual revenue and nearly $6 billion in annual operating profit. On average, these 15 companies are on a current trajectory (again, without any growth) to generate revenue of over $2 billion per year per company and operating profit of nearly $400 million per year per company. The simple signal that this small group sends is that there is a very large profit pool to attract the talent and appetites of adventurous executives in other industries.

These 15 solar power companies plus the 8 large producers of silicon feedstock are just the surface of a very large profit pool. These 23 companies are less than 0.05% of 50,000+ solar power players around the world. Given the many companies already in the sector, there will, of course, be storms of competition that eventually challenge these companies.

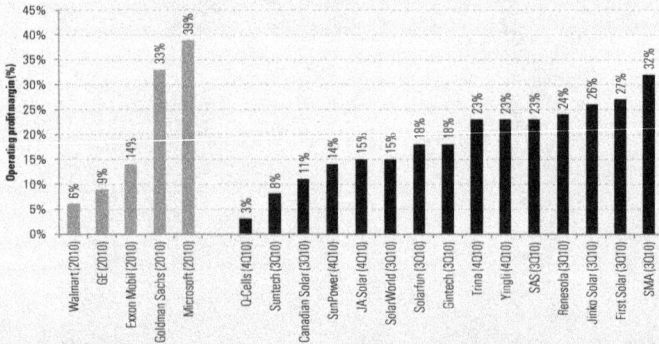

OPERATING PROFIT MARGIN »
(%)

KEY TAKE AWAY » Solar companies have profit levels approaching world's leading companies.

Yet, to date, these companies have demonstrated the resilience of their profit pool. One of the most impressive snapshots to prove this point is the image of solar power companies during the heart of the financial crisis. In the middle of 2009, when a broad set of companies were still going bankrupt, news headlines remained filled with negatives and the overall economy remained sluggish, there were many signs that the solar power sector was hurting.[7] For example, global average solar power module averaged $3.95 per watt in 2008 then declined by 42% to $2.30 per watt in 2009. This was a significant challenge for solar power companies. Prices declined 7% (minus $0.20 per watt) in the second quarter of 2009 and an additional 8% (minus another $0.20 per watt) in the third quarter of 2009.[8] This is a very fast price decline for a manufactured product. There was a general sense within the industry that the sky was falling and that doom was imminent.

However, most solar power companies exhibited remarkable resilience despite the macro-economic recession, illiquid capital markets and falling solar power prices. While the sky was falling for some higher-cost producers, there were many companies in this sector that sustained high operating profit margins. Despite challenges, there were a broad set of companies with operating profit margins above 10% in the third quarter of 2009, including First Solar, JA Solar, LDK, OCI, SMA, Solarfun, SolarWorld, Trina, Wacker and Yingli. The point is that a "normal" person would expect profit to go negative if (1) there is a major macro-economic shock, (2) financing for a product is at a standstill, (3) prices are dropping quickly and (4) there has recently been a large expansion of manufacturing capacity.

This "normal" mindset does not seem to be in play. In contrast to reasonable expectations, many solar power companies continued to thrive despite the challenges listed in the preceding paragraph. Many were profitable (as displayed in the following figure) and continued substantial capital expenditures for expansion. As a result, many grew quickly in 2009, including Canadian Solar (production +217% from 2008 to 2009), First Solar (+121%), JA Solar (+84%), LDK (+34%), OCI (+222%), Solarfun (+67%), SolarWorld (+48%), SunPower (+67%), Suntech (+42%), Trina (+98%), Wacker (+44%) and Yingli (+87%). These were growth rates of manufacturing firms in the middle of a deep recession! While there were problems for some solar power companies in 2009, the vast majority of companies

PROFIT MARGIN OF SELECT COMPANIES »
(OPERATING PROFIT %—3Q09)

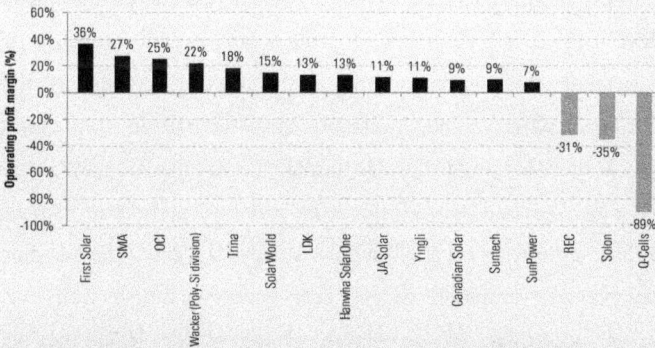

KEY TAKE AWAY » Solar power grew in 2009 despite the macro challenges because companies were highly profitable (albeit with some notable exceptions).

expanded and overall sector production (including inventories) increased from 7 billion watts in 2008 to 12.7 billion watts in 2009, equal to 80% year-on-year growth. In comparison to almost any other sector in the global economy, this sector demonstrated remarkable profit and growth. Doesn't it seem counterintuitive that a sector whose product is discretionary for end-customers would be able to thrive during a period of turbulence?

Yet looking at this situation from the perspective of negative network effects opens up a very different mindset. This is the mindset I am hoping CEOs and others reading this book will consider. As described earlier in the book, the prototype of a negative network effect starts with a "targeted attack on the best customers." If solar power companies were using the playbook of negative network effects, they would launch attacks on very attractive customers with the force and precision of Top Gun pilots. This is exactly their pattern. Solar power companies focus on customers that have very high willingness to pay for solar power. These customers are fairly easy to identify if you have the right map. The key is to find groups of customers that have a strong economic reason to adopt solar power.

If a customer gets a strong economic return from purchasing solar power, then the customer is likely to adopt solar power. This is not only common sense, but is also consistent with the data from my graduate school and consulting firm research. The next figure shows the economic return (customer internal rate of return, or IRR, in black) of customers adopting solar power in numerous different markets in 2010. This compares to the rate of return these customers might achieve investing in other vehicles (customer discount rate in gray). Although the mathematical maneuvering necessary to make these calculations is somewhat similar to the many moves a fighter pilot must make to carefully line up a clean shot, the result is obvious. Look at all the black above the gray! The customers to target for solar power are apparent because their economic returns from purchasing a solar power system far exceeds what they might achieve from other comparable investments. When you look at the many markets in which end-customer economics are very strong for solar power, it is easy to come to the conclusion that the world is a target-rich environment for solar power suppliers.

END-CUSTOMER RETURNS FROM INVESTING IN SOLAR POWER »
(%—2010)

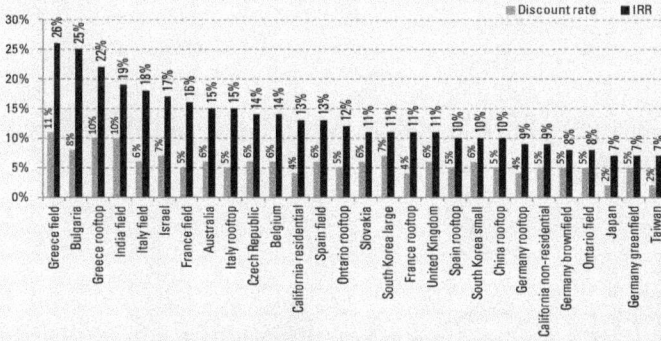

Chart legend: ■ Discount rate   ■ IRR

Categories (left to right): Greece field, Bulgaria, Greece rooftop, India field, Italy field, Israel, France field, Australia, Italy rooftop, Czech Republic, Belgium, California residential, Spain field, Ontario rooftop, Slovakia, South Korea large, France rooftop, United Kingdom, Spain rooftop, South Korea small, China rooftop, Germany rooftop, California non-residential, Germany brownfield, Ontario field, Japan, Germany greenfield, Taiwan

SOURCE: PHOTON Consulting, LLC.
NOTE: All data are rough estimates.

KEY TAKE AWAY » Many end-markets with attractive end-customer economics.

For example, the return from an Australian end-customer adopting solar power was 15% in 2010 without using any financial leverage. This end-customer return was 9%-points higher than the 6% rate of return for similar investments. The investment was more than twice as attractive. Guess what? "I've got radar lock!" is what the heads of Marketing yelled at they targeted Sales teams to capture these high-price customers. As a result, Australian solar power installations more than doubled from 75 million watts in 2009 to 190 million watts in 2010. According to one of Australia's largest solar power installers, "Growth in the PV business has been extraordinary." This appears accurate, with company solar power sales jumping >10X (!) from 2.4 million watts of installation in the second half of 2009 to 28 million watts in the second half of 2010.[9]

And the Australian situation is not unique. Similarly, Italian end-customers generated very high returns both from installing solar power on rooftops (15% IRR compared to 5% end-customer discount rate equal to a spread of 10% points) and from installing solar power in fields (18% IRR compared to 6% end-customer discount rate equal to a spread of 12% points). The result was stunningly rapid adoption of solar power both on rooftops (435 million watts in 2009 rising nearly 4X to 1.7 billion watts in 2010) and in fields (215 million watts in 2009 rising 6X to 1.3 billion watts in 2010). As in Australia, many companies outpaced the market average growth rates.

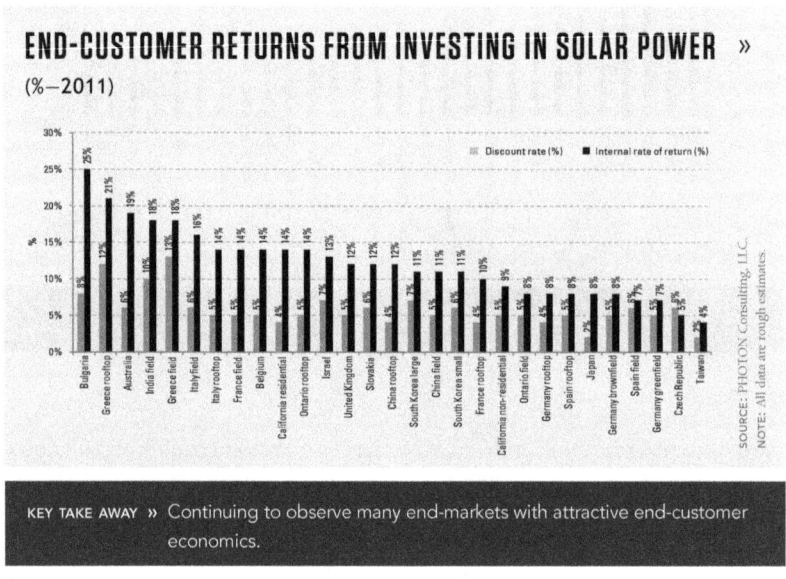

END-CUSTOMER RETURNS FROM INVESTING IN SOLAR POWER »
(%—2011)

KEY TAKE AWAY » Continuing to observe many end-markets with attractive end-customer economics.

The pattern is important. Solar power companies are able to launch sortie attacks on attractive end-customer market when end-customer returns are above discount rates. Whether 2X annual growth in Australia or 4X in Italian rooftops or 6X in Italian fields, the pattern is similar across markets and across types of customers. This pattern is important because it has held true for more than a decade. Markets in which the returns are strong display significant demand by end-customers to shift to solar power. In response, suppliers target the markets and customers in which strong customer economics enable high pricing for solar power systems and solar power modules. The suppliers simply pick among markets and customers in an effort to maximize their pricing and profit. For solar power companies, the good news remains apparent in the previous figure for

2011. Black (end-customer IRR) continues to be far above gray (end-customer discount rates) in a broad set of markets.

What the preceding paragraphs demonstrate is that the dynamic of solar power companies attacking the most attractive end-customer is very profitable and quite straightforward. But this dynamic is *not* normal. What makes this dynamic special is that solar power companies are on the positive side of a negative network effect. Whereas the USPS in the U.S. and the traditional electricity companies in Europe can see the downsides from negative network effects that reduce volume on their traditional infrastructure networks, the solar power companies can see only the upsides:

- **A large market with easy-to-find customers:** There are many groups of attractive end-customers who have economic desire to purchase solar power.

- **A very large profit pool:** The price that customers are willing to pay is far above the cost of solar power, which enables very high profit margins for solar companies despite increasing supply.

- **An even stronger value proposition:** The costs of delivering solar power continue to come down with scale and experience, enabling solar power companies to have conviction that the addressable market will continue to expand and that they will be able to defend high profit levels.

- **Small bumps in a downhill road:** Although there are challenges from traditional electricity companies in specific markets that will require shifting to other markets, these bumps appear unlikely to derail the gravitational force with which high returns in many markets accelerate end-customers toward solar power.

- ∘ **Even larger scale:** With demand obvious, profits high and alternative markets easy to identify, solar power companies have incentive to continue expanding rapidly, and so they do.

- ∘ **No end in sight:** Solar power remains well below 1% of global electricity consumption (approximately 0.3% in 2011),[10] making many-fold expansion plausible in the coming years.

The logic of solar power's growth is that it is on the positive side of a negative network effect, making it easy for solar power companies to focus on upsides. No single company embodies being on the positive side of a negative network effect more than the leading network equipment vendor in the solar power sector, SMA. Do you already know this company? SMA manufactures solar power inverters. Many may ask, "What is a solar power inverter?" Basically, solar modules generate direct current (DC), but most consumer electronics (e.g., televisions) use alternating current (AC). An inverter is a box used to convert DC to AC. The process of doing this is well understood (i.e., mid-tech not high-tech) and manufacturing an inverter is straightforward (i.e., low barriers to entry) with hundreds of companies making inverters today (i.e., many competitors) including behemoths like GE, Mitsubishi and Siemens.

If this were a "normal" sector, SMA would face very steep challenges, especially with a high cost structure compared to many competitors. Yet SMA has been able to grow very quickly and very profitably:

- ∘ Volume increased 23-fold from 344 million watts in 2005 to 7.8 billion watts in 2010.

- ∘ Revenue increased 10-fold from 172 million euro in 2005 to 1.8 billion euro in 2010.

o Operating profit increased 14-fold from 35 million euro in 2005 to 477 million euro in 2010.

SMA ANNUAL REVENUE[11] »
(EUR MILLION, %)

	2005	2006	2007	2008	2009	2010
Revenue (million euro)	172	193	327	682	934	1920
Operating profit (million euro—EBIT)	35	33	59	167	228	517
Operating profit margin (%—EBIT)	20%	17%	18%	25%	24%	27%
Revenue (million US dollars)	214	241	439	997	1299	2542
Operating profit (million US dollars—EBIT)	44	42	79	245	317	684
Operating profit margin (%—EBIT)	20%	17%	18%	25%	24%	27%
US$ per EUR	1.24	1.25	1.34	1.46	1.39	1.32

SOURCE: PHOTON Consulting, LLC, based on data from SMA.

KEY TAKE AWAY » SMA is leading example of company benefiting from the positive side of a negative network effect.

This company's performance has been outstanding. In 2009, during the heart of a global recession, the company grew revenue 37% year-on-year and achieved 24% operating profit margins. In 2010, SMA nearly doubled revenue and delivered 27% operating profit margins. This is a company with very high profit in a space with lower barriers to entry and many competitors. Further, its cost structure is high compared to its competitors. This description is just not normal. In a "normal" sector, SMA's high level of profit simply would either not exist or would quickly attract competitors. Yet SMA has been able to grow rapidly, achieve very high (40%) market share for solar power inverters and deliver high profit margins despite lower barriers to entry, many competitors and a relatively high cost structure.[12]

The success of SMA should inspire curiosity from any CEO who is looking to understand negative network effects (in general) and the solar power sector (specifically).

> *How did a start-up company with a high cost structure and no clear technological*
> *advantage come to control a large portion of the solar power inverter market and*
> *deliver very high profit margins from a commodity product in the face of competi-*
> *tion from behemoths like GE, Siemens, Mitsubishi and a sea of smaller players?*

Really, please think about this. It seems far-fetched, doesn't it? How the heck did this happen? How did SMA establish itself as the "Cisco of the solar power sector"?

The simple answer is: SMA understands network effects. It knows how to target customers on existing networks that purchase high price electricity and have a strong economic reason to switch to solar power. To capture these customers, SMA provides a reliable product and reliable service (training centers for installers, help desks for customers, etc.). But neither SMA's hardware (many other manufacturers are capable of producing hardware with similar performance at the same or lower cost) nor its service (many other companies are capable of providing a similar or higher level of customer service) is the true source of the company's differentiation. Its core differentiator from competitors (including very strong global conglomerates) is its ability to get to market faster than competitors. Its speed is enabled by a deep understanding of and a deep conviction in key market trends.

In a sector that grows as quickly as solar power, it is mission critical to identify markets before they turn on, to prepare tight logistics for products to reach those markets *before* demand is apparent and *then* execute on rapidly ramping supply to meet the demand. As shown in the previous black-and-gray figures of end-customer returns, the markets in which to focus can be easily identified. SMA simply does a better job than any of its competitors at identifying these markets, preparing to enter these markets and executing supply in these markets.

The speed of supply response is stunning. For example, the company expanded inverter shipments from 243 million watts in the first quarter of 2009 to 1.4 billion watts in the fourth quarter of 2009. This is 6-fold growth *within* a single year. By the third quarter of 2010, the company nearly doubled shipments *again* to 2.6

billion watts. **This type of explosive growth in shipments is only possible for a company that accurately predicts where the markets will be, carefully prepares channels to those markets, and then very rapidly executes on manufacturing and delivering its product.** No company could increase revenue 7-fold in 7 quarters (from 87 million euro in 1Q09 to 627 million euro in 3Q10) without being very good at forecasting. Having the financial and operational systems in place to manage such rapid expansion is impressive, and it is what makes SMA a special example of a company on the positive side of a negative network effect.[13]

SMA QUARTERLY INVERTER SHIPMENTS »
(MILLION WATTS/QUARTER)

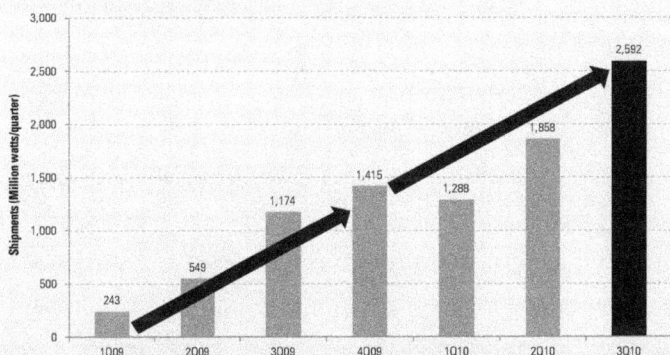

Shipments (Million watts/quarter)

1Q09	243
2Q09	549
3Q09	1,174
4Q09	1,415
1Q10	1,288
2Q10	1,958
3Q10	2,592

SOURCE: SMA quarterly reports and financial statements and PHOTON Consulting's *The Wall*. NOTE: All data are rough estimates.

KEY TAKE AWAY » SMA's shipments jumped 11 times from 1Q09 to 3Q10.

Clearly, there is a massive profit pool for companies that understand network effects and are capable of executing quickly. The scale of this profit pool is apparent throughout the preceding pages. It is also apparent that many companies see the pool and want to swim in it. In some cases, companies already benefiting from the profit pool realize that they may drown if competitors pass them with lower cost and faster growth. They realize that their portion of the profit pool may become more competitive. A combination of fear and greed leads these companies to actions that might be considered "strange" by a casual observer.

For example, take three companies that, on the surface, are manufacturing firms with deep expertise in semiconductors/electronics technology, but that made decisions to move into energy services and energy financing.

Cypress Semiconductor

Hoover's introduces Cypress Semiconductor by writing, "In Silicon Valley, it's perfectly logical for a giant Cypress to put its roots down in pure silicon."[14] It is the supplier of hundreds of different integrated circuits, programmable logic devices, clock and timing chips, USB microcontroller memory chips (including SRAMs) and other high-tech specialty products. Its customers include Cisco, EMC, Logitech and Sony. As Hoover's summarizes, "Overall, Cypress is a leading semiconductor design and manufacturing company."[15]

In the early 2000s, Cypress invested in a subsidiary, SunPower, to manufacture solar power modules. At that time, SunPower's cost structure was among the highest in the industry. Although its cost declined by 2006, SunPower's manufacturing process remained very high cost compared to its competitors. It feared that price competition would collapse its profit and reduce its growth. To prevent this, Cypress (the majority owner of SunPower at the time) made an unprecedented move for a semiconductor industry player: It acquired a leading solar power installation services company, Powerlight. This was a "strange" move because Cypress and SunPower's core strengths were designing and manufacturing technology products.[16]

The acquisition of Powerlight for more than $330 million was a bold move for a company with little skill in the energy sector (i.e., Cypress is a semiconductor company!), no skills in energy services (e.g., solar power installation services) and no skills in consumer and commercial financing (i.e., a requirement for solar power installations). Why did SunPower make this acquisition?

MEMC

According to Hoover's, "MEMC Electronic Materials doesn't waver from its devotion to wafers."[17] Historically, MEMC focused on supplying silicon wafers to leading semiconductor makers.[18]

Similar to the moves made by Cypress, MEMC is another leading semiconductor company that also did something strange. In mid-2009, this leading global producer of wafers for the semiconductor sector bought a solar power installation and financing company. Much like the Cypress-SunPower-Powerlight move, MEMC's move was "strange" because MEMC's core strengths were producing specialty chemicals and manufacturing high-tech feedstocks. The acquisition of SunEdison for nearly $300 million was a daring move by a company with little skill in the energy sector (i.e., MEMC is a semiconductor company!), no skills in energy services (e.g., solar power installation services) and no skills in consumer and commercial financing (i.e., a requirement for solar power installations). Why did MEMC make this acquisition?

Sharp

Hoovers describes Sharp's business as, ". . . pointed at the electronics market. Best known for its consumer electronics, the company is also a leading maker of electronic components and computer hardware and peripherals." Its core makes LCDs, flash memory, integrated circuits and laser diodes. Plus PCs, printers, cell phones, consumer audio and video products and a variety of appliances.[19] A minor part of Sharp's business has focused for decades on manufacturing solar power modules, but not in solar power services or financing.

Guess what? Sharp followed in the footsteps of Cypress and MEMC. In 2010, Sharp acquired U.S. solar power installation and finance player Recurrent Energy. The acquisition of Recurrent for $300 million was a bold move by a company without deep experience in the energy sector (i.e., Sharp is an electronics manufacturer!), no skills in energy services (e.g., solar power installation services) and no skills in consumer and commercial financing (i.e., a requirement for solar power installations). Why did Sharp make this acquisition?

Why did these companies make these moves? These are not business-as-usual moves. Jumping from semiconductor/electronics technology and manufacturing into energy services and financing is *not* normal. In 20 years of experience in the energy sector, I have not observed companies without real strength in energy acquiring energy companies. Similarly, in nearly two decades of experience working with Asian conglomerates, I have never observed a foreign acquisition that is so far from the core business as Sharp's acquisition of Recurrent.

These are *strange* moves that indicate both *greed* and *fear* are involved. These companies can see the massive profit pool. Not only can they see it, but each has dipped its toe into the pool and has achieved noteworthy success in starting a solar power business. All exhibit signs of *greed* in trying to become the leading solar power company. Yet all three have come face-to-face with the realization that they may not be able to compete with lower cost, faster moving rivals if they continue operating only as manufacturers. Out of *fear*, they moved from manufacturing into energy services and financing. These moves made their manufacturing operations *safer* by ensuring internal demand for their manufactured products and reduced time-to-market (something SMA has mastered but other solar companies have not).

Overall, these companies have realized that capturing the positive sides of negative network effects involves looking at the world in a different way. They are not alone. Other solar power manufacturing companies have made the jump into energy services and financings (e.g., silicon maker OCI's acquisition of U.S. electricity player CornerStone Power Development, solar manufacturing equipment vendor AEG's acquisition of Spanish solar power installation specialist OpcionDos). And many more solar power companies are in the process of similar moves. Further, my core hypothesis is that this is *not* just solar power and *not* just electricity, but a broader trend toward distributed infrastructure that appears likely to have broad reaching implications (both positive and negative) for energy, food, water and communications infrastructure.

Whether driven by greed or fear, the two main questions for any executive who wants to *really* understand negative network effects is, "Where is value?" and "How to capture it?" These are the central questions for the chapters that follow. The main point of this chapter has been to try to explain why others might be greedy about the possible upsides of negative network effects when I am fearful of the downsides. My hope is that there will be leaders to heroically run after this storyline. Like Rocky hitting an opponent with a combination-punch barrage to win a match or like Indiana Jones running into an impossible situation in search of eternal glory, my hope is that the stage is set for the large profit pools created by the positive sides of negative network effects to attract top management talent.

Chapter 5

FINE CHEFS WITH
EXPLOSIVE INGREDIENTS

*The credit belongs to the man who is actually in the arena
. . . so that his place shall never be with those cold and timid
souls who neither know victory nor defeat.*
—Theodore Roosevelt (1858–1919)

EXPLOSIVE INGREDIENTS

Lucky me. I am not a good cook, but I married into a family of chefs. Susan and members of her immediate family have cooked in restaurants, been food critics, graduated from top culinary schools, published cookbooks, started food magazines and owned restaurants. Before meeting her family, I wasn't aware how much effort chefs make to ensure that they have high quality ingredients before starting to cook a meal. It makes sense that the quality of the outputs depends on the quality of the inputs. The menu of Chef Thomas Keller's French Laundry, for example, uses hand-selected ingredients in each dish. For example, on a recent evening, the "peas and carrots" dish is composed of grilled hearts of palm, Akita Komachi rice, sugar snap peas, carrots and Madras Curry "Aigre-Doux." Making such fine cuisine requires exceptional ingredients in the hands of exceptional chefs.

FINE CUISINE »

SOURCE: PHOTON Consulting, LLC.

KEY TAKE AWAY » Fine cuisine requires exceptional ingredients in the hands of exceptional chefs.

This perspective is important for CEOs pondering a world filled with negative network effects and asking the central questions, "Where is value?" and "How to capture it?" Answering these questions may require new skills, even for large existing companies, which may require new recruiting, even for companies with deep management talent. Being on the profitable side of negative network effects is a significant effort hinging on the CEO's honest evaluation and deep commitment. **The central message of this chapter is that CEOs (much like top chefs planning fine meals) should carefully check availability of high quality ingredients necessary for high quality results. This chapter presents a real-world example (i.e., the food industry) of how these ingredients help a CEO see opportunities that others miss. Basically, without high quality inputs, the outputs may not turn out as planned.**

Almost all companies in nearly all industries have processes for strategy, finance, procurement, operations, marketing, sales, research & development (R&D) and human resources (HR). These skills may be necessary, but they do not offer much hope of differentiation in sectors benefiting from negative network effects. Sectors exhibiting "explosive growth" on the "positive side of negative network effects" benefit from very fast uptake because their product substitutes for an existing network's offering. In this dynamic, companies that know the markets best are able to act fastest, which enables them to capture higher profit with lower risk. To do this successfully, companies need more than typical "marketing" skills. They need skills that have a significant element of "strategic marketing." These companies often use strategic marketing skills as a key tool to inform strategic choices about (1) geographic markets, (2) segments within specific geographic markets and (3) specific customer needs within specific segments.[1]

The next chapter will build a 1,000 day plan for industries on the positive side of negative network effects. However, before building a plan, **this chapter will focus on the important human resource ingredients necessary to truly pursue the positive sides of negative network effects.** Focusing on these ingredients is important *before* starting to plan the menu or cook the meal because not many companies have a stock room filled with the talents and personalities necessary for "strategic marketing" to be effective. How many companies possess a corporate executive who combines the traits of Sherlock Holmes, Indiana Jones and Rocky Balboa? Very few. Even companies that have very strong traditional marketing (e.g., P&G, Coke, McDonalds, Nike) may find their strategic marketing talent in short supply.

CEOs seeking to participate in the explosive growth opportunities generated by negative network effects must *first* evaluate if their companies have a peculiar set of skills and traits necessary to evaluate and then act. Observation of thousands of executives in the solar power sector has led to a list of 20 personality traits and skills necessary for companies to successfully build strategic marketing in sectors benefiting from negative network effects. This long list includes leaders who:

1. Are flexible thinkers;
2. Have remarkable lack of ego in that they are more interested in getting to the right answer instead of being right themselves;
3. Have deep curiosity on a broad set of topics;
4. Have both big picture vision and attention to detail focus;
5. Are willing to leverage other people's ideas;
6. Are willing to adjust their own ideas and admit their initial answer is wrong if a better answer emerges;
7. Defend their views fiercely in the absence of stronger logic;
8. Have precise observation and detailed memory;
9. Have strong analytical aptitudes;
10. Have tight reasoning capabilities;
11. Are inspirational written and oral communicators;
12. Interact flexibly with a diverse set of personalities;
13. Understand how to operate in complex multinational business environments;
14. Understand how to operate business in which public policy plays an important role;

15. Are slightly more interested in impact than the right answer, but still care deeply about the right answer;

16. Move quickly through the progression from "unaware" to "aware" to "understand" to "believe" to "act";

17. Have been professionally successful *enough* to have significant responsibility and true authority in business settings, while not being so successful that they have lost even a degree of internal drive;

18. Have unflinching commitment to succeed despite challenges;

19. Have deep personal integrity; and

20. Put the needs and feelings of others ahead of their own.

These 20 characteristics are a list of ingredients. In cooking, using the right recipe with the wrong ingredients is unlikely to lead to success. Similarly, building a 1,000 day plan without the right "strategic marketing" ingredients of personality and skills is unlikely to be successful. This chapter focuses on the mission-critical task of ensuring that CEOs recruit the right ingredients before they begin implementing a strategic marketing plan.

A CEO might look at this list and think it too long. This *is* a long list, but these personality traits and skills are essential because they fit the challenge at hand. Basically, a leader in a period of negative network effects must be able to act well before all necessary information is available and as a result must perpetually seek to improve the answer and revise the actions. Doing this requires a high level of communication skills and deep trust by colleagues of the leader's personal and professional character.

Given that this list of 20 is far from a succinct Jack Welch-ian business aphorism (e.g., "Number one, cash is king . . . number two, communicate . . . number three, buy or bury the competition."), it is important to make the point more tangible. So let me share with you the strongest personification of these skills and traits that I have found: Frank Britt.

Frank grew up in a diverse and economically modest town bordering the Bronx. He never lacked life essentials and parental encouragement, but his main mode of operation was trying to stay one step ahead of trouble. He became effective at navigating to avoid at-risk situations that surrounded him. Along the way he started to see how even small life choices can define a person's future. To him, it was clear that being at the wrong place at the wrong time could have significant and devastating long-term consequences. Fortunately, he had conviction that a better life was achievable through hard work combined with some "earned" luck and supportive mentors. By age 16, Frank was actively seeking advice from accomplished adults and committing himself to the hard work of getting on a path that he, and others in his family and his neighborhood had not taken. This path included Syracuse University, which he convinced to take a chance on him based on his determination and natural talent, rather than demonstrated GPA. As the first member of his family on this path, Frank moved from a potential blue collar trajectory into the world of the college educated. Unlike many of his classmates from more privileged backgrounds, Frank had seen firsthand the limits of career opportunities at the lower rungs of the socioeconomic ladder and realized the strong connection between good grades, college graduation and long-term income potential.

With relentless determination, Frank has held on to opportunities with both hands. He excelled academically and not only graduated among the top of his management school class, he was recognized as the top undergraduate logistics student in the United States. He received supply chain management and marketing job offers from over 20 Fortune 500 firms such as Pepsi, IBM, Ford and Andersen Consulting.

Over the last 20-plus years, he has made steady progress up the professional ranks. His first job out of college was in the operations management practice at Arthur D. Little (ADL), at the time, a 102-year-old management consulting firm and arguably the original business think tank. This was a major achievement in itself because no one from his school had ever been recruited to ADL, a firm flush with elite professionals from the top schools, including a workforce comprised of 60% PhDs. He was the first. He spent nearly a decade climbing the corporate ladder as a management consultant at ADL (to the role of senior analyst), then moved to global consulting giant Accenture (to the role of senior manager), then joined as a vice president an early stage Internet retailer he helped build and successfully took public, and finally joined e-business consulting firm Mainspring (where he led the retail and consumer products group). IBM acquired Mainspring in 2001 as the dotcom bubble burst, Frank's ability to stay one step ahead of trouble continued, and he was soon running a $400 million division of IBM.

By his mid-30s, Frank was already a vice president of consumer products responsible for a $400-million division within one of the world's largest companies. Yet he was constrained by the corporate culture of IBM and boldly sought other opportunities to "bulk up" his operational capabilities and expand his horizons. Leveraging his expertise in IT outsourcing and his experience with retail and consumer products sectors, he joined C&S Wholesale Grocers (currently the 10th largest privately held company in the U.S. according to Forbes) as the vice president of supply chain management. In this role, he was accountable for managing the core business process of product fulfillment for 12 major U.S. grocery retailers such as Stop & Shop and Safeway, including over 4,000 stores. This required optimizing a complex value chain of over $700 million in finished goods inventory and 250,000 stock-keeping units sourced from over 1,000 vendors and staged across dozens of company warehouses.

By this point in his career, Frank was publicly signaling his strategic marketing capabilities. For example, *Super Market News* quotes him during this period saying, "Everyone is washed with data and struggling to make sense of it, but the companies who are really getting out in front do a much better job of linking transactional information and shopper insights."[2] This is the vice president of supply chain management of an old school company speaking directly to the imperative of competing based on superior analytics, instead of only scale, and highlighting the power of advanced analysis in order to make better fact-based decisions with strategic implications.

By the mid-2000s, his professional career had been successful enough, but he was still hugely hungry for further achievement not possible within C&S. With his deep curiosity on a broad set of topics, Frank set out to find a larger professional adventure. In 2007, he jumped to a totally different industry and job function. He joined Pri-Med, a leading provider of medical education for health care professionals with the backing of Bain Capital, as general manager of U.S. operations. He was soon named president and CEO. Frank took the control during the 2009 downturn when the company was in steady decline and in risk of failure due to significant market and regulatory shifts. He and his team re-created the company's business model and cost structure and transformed it into a leaner, profitable and growing company. This is where he stands today. Frank is at the helm of a company that manages a network of 330,000 health care professionals (including nearly 50% of all primary care physicians in the U.S.) and influences over 140 million patients annually. It will be fascinating to see where he goes from here.

I met Frank about 5 years ago, just before he joined Pri-Med. We were introduced via email by Shawn, the CEO of a hedge fund who intuitively knew that I would be thrilled to meet someone who has deep strategic marketing insights. Shawn knew I would value give-and-take conversations with Frank on a broad variety of topics. Shawn was emphatic about the meeting. He even wrote the two of us an email titled, "Getting you together" in which he insisted that we meet. "I'll even spring for the meal just for the pleasure in knowing that you two have finally met."

From the start, I both admired Frank and was fascinated by Frank. To me, Frank is the perfect combination of curious 10-year-old who is captivated by everything on the planet and an aggressive 40-year-old who is hungry to make an impact on the business world. You might say Frank is a fighter with heavyweight champion caliber. Frank's heavyweight side includes knockout features: Relentless energy, incredible intellectual absorptive capacity, lightning fast analytical skills, heart-warming charisma and steadfast loyalty, among many others.

And guess what? He wins. Frank has built a stunning career record with many knockouts that includes: First member of his family to graduate with a four-year college degree, first member of his college to be recruited into Arthur D. Little, a senior manager at a leading consulting firm by ~30, vice president responsible for hundreds of millions of dollars by ~35, president and CEO of a company by ~40. Even more importantly, Frank has amassed 20+ years of impressive professional achievements while building an incredibly loving, balanced and fun family, and also contributing to his communities in countless impactful ways. By nearly any measure, Frank is a heavyweight.

Why have the last three pages been dedicated to convincing you that Frank is a heavyweight? Because you, the reader, are the equivalent of a customer in a fine restaurant. The chef has spent many years practicing and many hours preparing for you to walk into the restaurant and order from the menu. To convince you of the quality of the meal even before you order, the chef signals to you that the ingredients have been hand selected. This process of convincing you about inputs is necessary because you might not otherwise trust the outputs. Remember, Frank is a successful CEO in the health care industry with senior executive experience in the technology, education, food and retail sectors and a decade-plus of consulting experience.

Tonight's menu?

An assessment of the centralized food industry with the key ingredient "insights about negative network effects" made with a 5-year aging process of monthly lunch and weekly email exchanges with Frank in which he has been a thought partner, coach, mentor and mental sparring partner.

The seeds for this meal were planted by Frank's wife Noreen. She sent Frank a chart describing the growth of farmers markets in the U.S. from 1994 to 2010. During this period, farmers markets expanded from 1,755 to 6,132, equivalent to 8% compound annual growth. In 2010, the growth rate was 16% year-on-year, indicating an acceleration of the trend.[3] Frank knows that I am interested in data that might support or refute potential negative network effects in the food sector, so he sent the link to me as an FYI. We then went back-and-forth in a long email exchange. Before I share the results of that email exchange, please take a close look at the data. Really look at the data and ask yourself, "Are the seeds here for a negative network effect?" Then keep reading.

FARMERS MARKETS IN THE US 1994–2010 »
(NUMBER OF FARMERS MARKETS)

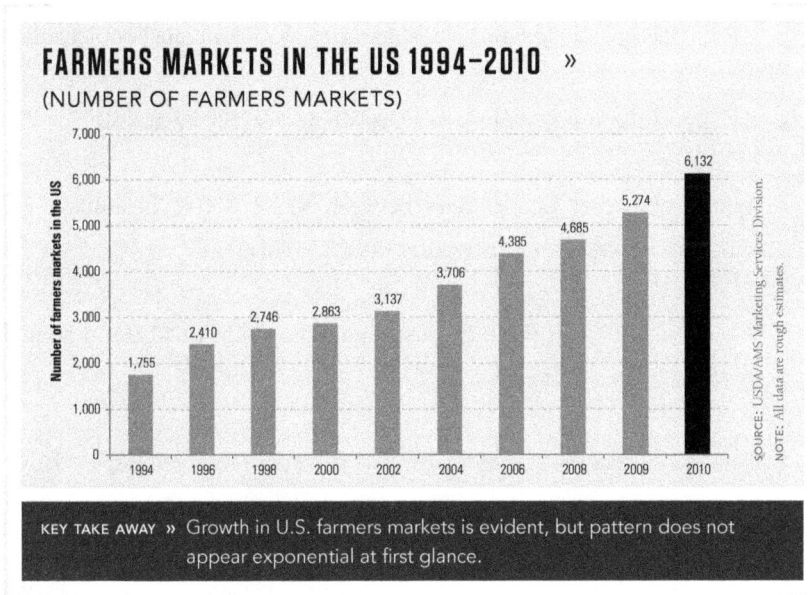

KEY TAKE AWAY » Growth in U.S. farmers markets is evident, but pattern does not appear exponential at first glance.

At first, Frank and I both thought that these data were *not* indicative of a negative network effect. For example, Frank's initial reaction to the figure was similar to my initial thinking. Specifically this is a section from email Frank sent to me based on his experience in the retail and consumer packaged goods industry. The purpose of sharing this unsynthesized email dialogue is to demonstrate the quantified banter that takes place as Frank searches for the right answer to a complex problem.

- Total U.S. produce sales at retail ~$120B including $60B in grocery channels, etc.

- 28k grocery stores (not including supercenters) = 70% of retail sales are grocery stores (the remainder are supercenters).

- I saw data that said total of $2B farm stands and public markets (very important point: The remainder would be in food service channels).

- $2B divided by 6,000 farm stands/markets = $333k in retail per location.

- So, if you add 6,000 more farm stands/markets you are at $4B.

- Typical a grocery retailer does between $25M–$50M in annual total sales (btw, a Walmart Supercenter does $100M+).

- Assume 25–30% in produce (just a guess) or $6M to $12M in annual produce sales.

- I am not sure of the size of fresh cut as percent of total retail sales. However, we do know fresh produce consumption is growing versus most categories and that the gross margins are higher for produce and that the profitability of a typical grocery store is between 3% and 5%. We also know "going local" is a mega-trend in grocery retailing in general (including Walmart that has changed its procurement efforts in this area radically in the past 2 to 3 years to appeal to the growing consumer sensibility on local and fresh).

- SO, the question would be: "IF grocery chains lost 20% of the highest margin demand (and a key destination category for consumers) it would materially impact profitability and compel further in-store cost cutting. Is that plausible?"

- In the short term: NO. I think it is a stretch to suggest this would happen anytime soon. If a grocery store is doing $6M, can a farm stand model siphon off $1.2 M in a local market within a 4 to 6 mile trading radius? I doubt it.

- Moreover, what likely really happens (I am guessing) is the total category sales grow as people are inclined to consume more rather than substitute. Within reason, there is an expandable consumption aspect to the category, aka, I eat more bananas when I have bananas.

- If it COULD happen over the next decade, however, it would clearly be a very localized impact on a store specific basis.

Frank's initial conclusion was similar to mine: Farmers markets appear unlikely to cause a negative network effect that impacts the network of centralized food including grocery stores. While we both thought it could occur, we were none-theless convinced that it would be very store specific and highly localized, not a macro-trend.

But then we kept looking at the data. And we started to see different trends as we looked more closely. What we learned is that it doesn't take much to stop growth at the local level in a traditional food network. More specifically:

- If an average grocery store has between ~$7 million and ~$16 million in revenue (65,000 supermarkets with ~$465 billion in revenue = ~$7 million/store)[4]

- Let's say that an average grocery store has a 3% pre-tax margin (this may be high given that TESCO and Walmart have ~5%–6% pre-tax margins for their overall retail business, and there are many retailers with margins below this)

- This equals pre-tax profit of $215,000–$469,000 per year (3% of ~$7.2–15.6 million)

- Now let's say that store revenue increases 5% to ~$7.5–$16.4 million AND pre-tax margin expands from 3% to 4%, so that pre-tax profit is ~$300,000–$657,000

- This basically means that pre-tax profit of the store increased by ~$85,000 (~$300K – ~$215,000) to ~$188,000 (~$657,000 – ~$469,000)

Frank had initially posed the question, "If a grocery store is doing $6M can a farm stand model siphon off $1.2 M in a local market within a 4 to 6 mile trading radius?" His initial conclusion: "I doubt it." Me too.

But then we started to reconsider. Perhaps you don't need to get a $1.2 million revenue reduction to cause a problem because the profit side will be an issue first. As with solar power, if decentralized food causes questions to be raised about the volume, price and cost of centralized food, then there may be a *big* reset button for the entire system *much* faster than the $1.2 million per store revenue loss. In addition, the impact may be broader because capital markets may stop financing centralized food assets without much higher returns.

Perhaps a better question is, "How much revenue from farmers markets is enough to stop *profit* growth for an average store?" This is the central question. If distributed food stops the profit growth of centralized food, then there may be *big* problems. And it doesn't have to stop the growth of profit for all centralized food, but only for centralized food where wealthy people (the same people who disproportionately support the profit of the food sector) live.

Our guess-answer to the question is ~4 average-size farmers markets per average supermarket is enough to eliminate all profit growth for the supermarket. Here are the basic assumptions:

○ A farmers market disproportionately impacts the sales of nearby supermarkets.

○ More specifically, the impact is even more disproportionate for produce and other items that are being sold at farmers markets.

○ One supermarket store covers one zip code. (Note: There are between 36,000 and 65,000 supermarkets in the U.S. located in 43,000 zip codes. Assuming just over 300 million people in U.S. and 43,000 zip codes, this equals 7,000 people/zip code. With 7,000 people and an average of $2000 per capita spending on groceries, this equates to around

$14 million/supermarket. This is probably a bit high because some of the spending is via convenience stores, etc.)

o Revenue is $7–$17 million/supermarket.

o One farmers market in the zip code with $350K in revenue per market per year.

o For the point of this exercise, the assumption is that 100% of the $350,000 revenue is from people within the zip code. A more accurate estimate might be that the $350,000 in revenue is split 50–50 between 50% people from the zip code who do their shopping in the zip code and 50% people from outside the zip code who do their supermarket shopping outside the zip code of the farmers market. However, farmers markets cluster and other nearby farmers markets might take an additional 25% or more from the supermarket's zip code, so it is harder to estimate.

o Nearly all of the volume purchased at the farmers market displaces volume that would be purchased from a supermarket, but the revenue displaced is lower (i.e., price at the farmers market is 20% higher than the supermarket price for same item), so only ~$292,000 of supermarket revenue is displaced.

o The average pre-tax margin for the supermarket is 4% but produce (especially high-end organic produce) is 2X the normal margin . . . so 8%. As a result, $23,000 of pre-tax profit is displaced.

o The result is that in a zip code that has lower revenue/supermarket (i.e., ~$7 million/supermarket), it would take ~4 farmers markets to completely eliminate profit growth. For a zip code with higher revenue/supermarket (i.e., ~$16 million/store), it would take ~8 farmers markets.

○ Overall, this analysis suggests that you might begin to see challenges for traditional supermarkets starting ~4 farmers market/supermarket.

Does 4 farmers markets per supermarket seem like a lot? It did to us. But then we looked at some data on Boston. Many of the zip codes in Boston range from 02100 to 02199. Here is a figure showing the zip codes in/around Boston with farmers markets registered with the USDA. Dark gray in the figure indicates more than two registered farmers market per zip code. Black in the figure indicates three farmers markets per zip code. There are *already* several zip examples of geographies (zip codes) with 2–3 farmers markets. Pretty quickly it became apparent that 4 farmers markets in a small geography might be more plausible than we had thought.

FARMERS MARKETS REGISTERED IN ZIP CODES 02100-02199 »
(NUMBER OF REGISTERED FARMERS MARKETS)

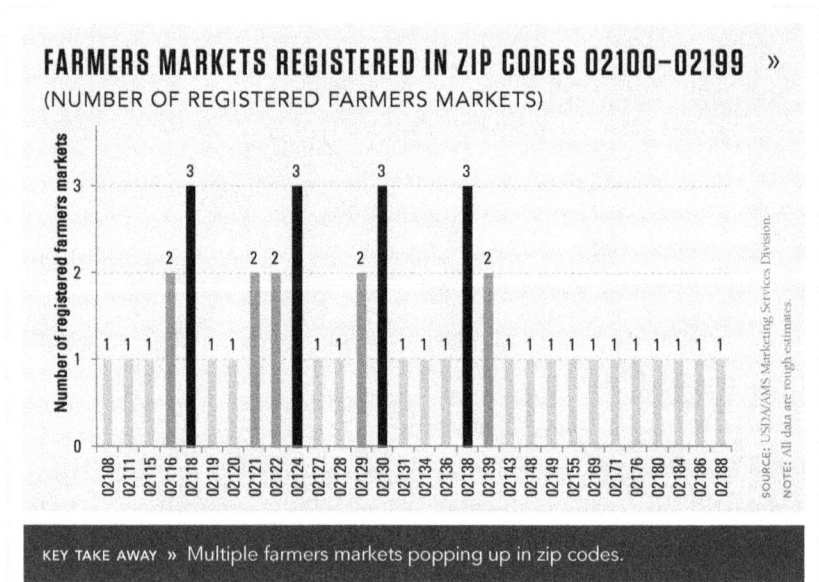

SOURCE: USDA/AMS Marketing Services Division.
NOTE: All data are rough estimates.

KEY TAKE AWAY » Multiple farmers markets popping up in zip codes.

Initially we thought, "The Back Bay and South End of Boston are not representative of the world." But then we realized that they are wonderfully representative of the wealthy zip codes around the country: Dense population, over-served with centralized food, increasingly served by distributed food (including farmers markets), high cost to serve and high price for the customer. And, perhaps most important, these areas contain the generators of disproportionate profit in the centralized food system. It's not just that these customers are higher profit, it's that serving these customers makes it easier for centralized food networks to serve other somewhat similar (geographic, demographic, etc.) customers. If you take away the most profitable customers, stores and product volume, the rest of the network will face a big challenge making up cash flow, especially when they need to convince financiers to provide more debt or more equity. And newer farmers markets seem to be highly focused on this type of geographic/demographic area.

An initial review of data points to this. As displayed in the following table, urban areas like Back Bay-South End (Code 1) are 36% of all farmers markets created in the last 5 years including 68% of all large farmers markets (>$23,000/month revenue). If you add the contiguous regions with slightly less density (Urban Code 2), then you get 50% of total markets (36% + 14%) and 84% (68 +

FARMERS MARKET BY RURAL-URBAN CODE AND BY SIZE OF MONTHLY SALES[5] » (%)

Rural-urban continuum code	U.S	Percentage of Monthly Sales			
		Less than $2,500	$2,500 to $6,999	$7,000 to $22,999	$23,000 or greater
1	35.6	28.6	22.2	31.8	68.0
2	14.0	8.2	13.9	11.1	16.0
3	14.3	18.4	8.3	31.8	0.0
4	7.0	10.2	8.3	31.8	8.0
5	5.5	4.1	11.1	9.1	8.0
6	11.1	16.3	16.7	6.8	0.0
7	5.0	0.0	5.6	2.3	0.0
8	3.8	10.2	5.6	0.0	0.0
9	3.8	4.1	8.3	0.0	0.0

SOURCE: USDA/AMS Marketing Services Division.

KEY TAKE AWAY » Farmers markets appear to be a focused attack on high-priced customers within centralized food networks.

16%) of the largest markets. Much like what has occurred with distributed solar power attacking centralized electricity networks and quickly capturing customers or online bill pay attacking centralized mail distribution networks and quickly capturing customers, these farmers markets seem positioned for a pretty focused attack on centralized food networks.

The main thing to share from my academic and consulting firm research over the last decade is this: A negative network effect that financial markets care about normally occurs when (1) an exponentially growing number of customers (2) in a dense geographic region begin turning to a different network for (3) high price products (4) to which they say they are not very price sensitive (as long as the IRR from buying the product is above the customers' discount rate). The occurrence of a negative network effect in one specific network may be (5) further augmented by defections of existing customers across other similar of products (i.e., there may be multiple negative network effects occurring at once). When these five things happen, you have basically undercut the core reasons that financial markets believe the existing network's volume, price and costs are reliable.

Here is a discussion of what Frank and I are finding as we dig deeper into the topic.

1. An exponentially growing number of customers

When Frank first sent the data, we looked at the cumulative number of farmers markets. Below is a redrawn representation of the data shared in a previous figure after extrapolating to make estimates for all years 1994 to 2010. It shows a regular upward trend, but the growth rate is only 8% CAGR 1994–2010. This (~2X GDP growth) seems too slow to be worthy of big notice. Even the most recent years (2009, 2010) show only 14% CAGR 2008–2010. Yes, 2010 is the fastest annual growth rate on record, but the 2008–2010 growth rate (14% CAGR) was actually slower than the 1994–1996 rate (17% CAGR). In order to really capture attention, we would want to see compound annual growth well above 20% and possibly well above 30% per year.

FARMERS MARKETS IN THE UNITED STATES, 1994-2010 »
(OVERVIEW)

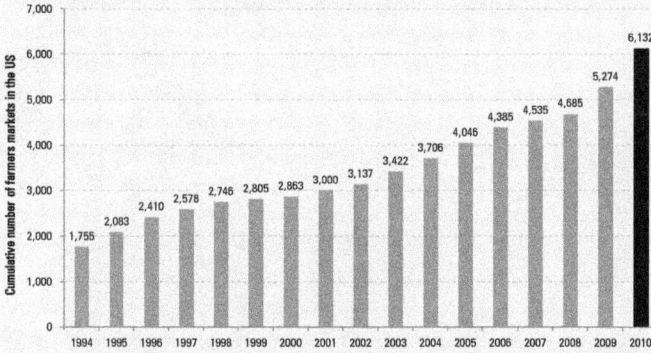

KEY TAKE AWAY » Initial analysis suggested linear not exponential growth.

The problem with looking at the installed base is that you want to look at growth in "growth terms" not in "growth of cumulative base" terms. Take a look at the same data as shown in the previous figure, which now shows the number of *new* farmers markets per year instead of just looking at the total number of farmers markets per year. The next figure is a representation of the data after extrapolating to make estimates for all years from 1994 to 2010. It's a very large jump from 2008 to 2010. A best guess after extrapolating the data is that new farmers markets in 2010 were nearly 6X larger than new farmers markets in 2008. Even compared with the largest growth years in history (e.g., 328–340 per year in 1995–1996 and 2005–2006), the years 2009 (589 new) and 2010 (858 new) look remarkably strong. This seems to be a key metric to follow. If we keep seeing new farmers markets grow quickly, then (of course) the total number of farmers markets will grow quickly. This growth in *new* farmers markets is the number-one lead indicator as we ask the question, "Is a negative network effect likely in centralized food due to the shift of volume to other food networks?"

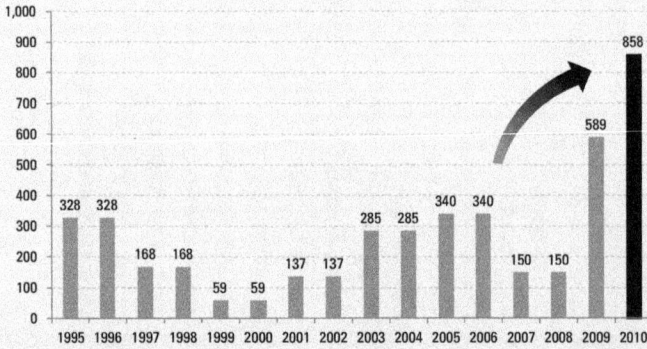

NEW FARMERS MARKETS IN THE UNITED STATES, 1994–2010 »
(NUMBER OF ANNUAL ADDITIONS)

SOURCE: PHOTON Consulting, LLC, based on data from Agricultural Marketing Service of the U.S. Department of Agriculture

NOTE: All data are rough estimates.

KEY TAKE AWAY » Further analysis suggests exponential growth.

In addition to the number of new markets, we are also interested in the growth of customers at existing markets. Anecdotal evidence suggests that there is growth in customers/market. Susan tells me this from her experience at the farmers markets near our house and from the farmers markets pretty far from our house to which she travels regularly. "They are all bigger now than the year before than the year before than the year before . . ." Beyond her personal experience, there are also reports on farmers market. The next figure is from a report on the Union Square Farmers Market in Somerville, Massachusetts. It shows that the weekly attendance jumped a lot in 2010 compared with previous years. This would indicate that there are not only more new farmers markets each year, but that existing markets have more attendees. All of which signals that negative network effects may be much closer at hand than many people would think. We still have more homework to do (i.e., these are just anecdotes) and we still need more time to pass (i.e., "Will these trends continue?"), but this seems to ring true so far from interviews/analysis/anecdotes. We are looking for counterexamples, but not yet finding them.

MARKET AVERAGE WEEKLY ATTENDANCE AT UNION SQUARE »
(NUMBER OF ATTENDEES/WEEK)

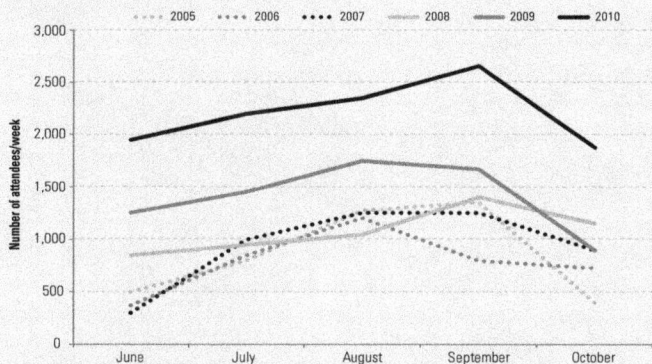

KEY TAKE AWAY » Steady increase in attendance at many farmers markets.

2. Dense geographic region

The Agricultural Marketing Service of the U.S. Department of Agriculture (USDA) has developed a map of Farmers markets. The map shown in the next figure illustrates clear clustering. We have done some preliminary analysis of the specific zip codes and are in contact with the people who track the data on farmers markets at USDA. An initial review (we still need to look more carefully toward finding problems with the hypothesis) suggests that the new farmers markets are occurring disproportionately in or near neighborhoods that already have farmers markets. Basically, "nothing attracts a crowd like a crowd." A geographic region that has one farmers market is disproportionately more likely to have another... and another... and another... The result (that we need to prove more with statistics but can already see in initial review of the data) is that the cumulative markets trend that jumps out in the previous figure (farmers markets registered in zip codes 02100 to 02199) becomes even more jarring when you look at the number of new markets in a specific cluster of zip codes.

FARMERS MARKETS IN THE UNITED STATES »
(MAP)

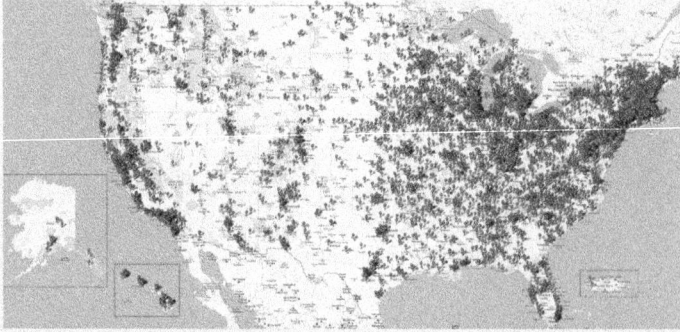

SOURCE: Agricultural Marketing Service of the U.S. Department of Agriculture.
NOTE: All data are rough estimates.

KEY TAKE AWAY » New farmers markets disproportionately near existing farmers markets.

For example, take a closer look at Massachusetts on this map. It's like an ant farm crawling with ants. The farmers markets are so dense that you can't even see the underlying map. This density is apparent to anyone (like Frank and me) in Back Bay over the last ~5 years. The area has gone from one farmers market one day per week to two days per week (Tuesday & Friday in Copley Square) and another farmers market (Prudential) on an additional day per week (Thursday).

The density of a typical farmers market is striking. Look at how >50% of the Union Square Farmers market draws its customers from a single zip code (02143) and draws nearly all remaining customers (>80% of total) from nearby zip codes. Wow, this is dense geography. This is similar to the distribution of zip codes for a center-of-suburban-town gasoline station. Interesting. The point of geographic density is that once you prove that the economics of a distributed network are solid, then a "pile-on" effect seems to occur. A full parking lot, lines at stands, or 100% of the stalls filled with farmers/merchants makes it easy to see that a farmers market is attracting too many people. This is normally when the next farmers

FARMERS MARKETS IN THE NORTHEAST UNITED STATES »
(MAP)

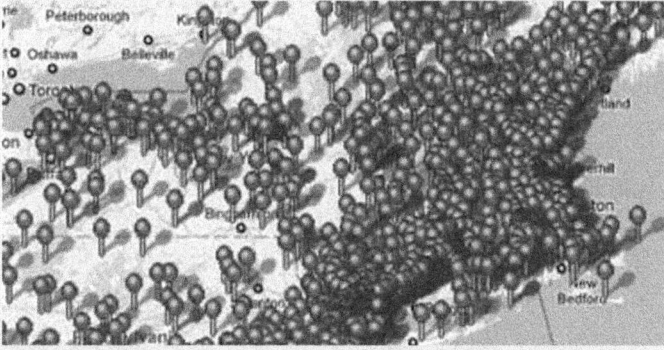

SOURCE: Agricultural Marketing Service of the U.S. Department of Agriculture.
NOTE: All data are rough estimates.

KEY TAKE AWAY » Nothing attracts a crowd like a crowd.

RESIDENCE ZIP CODE OF UNION SQUARE FARMERS MARKET CUSTOMERS IN 2010 » (% BY ZIP CODE)

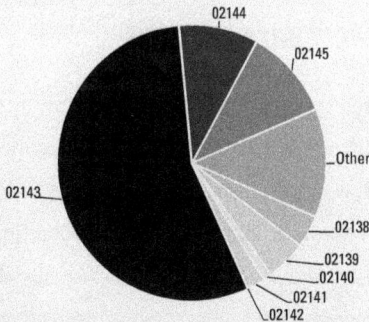

02144
02145
Other
02138
02139
02140
02141
02142
02143

SOURCE: PHOTON Consulting, LLC, based on data from "Union Square Farmers Market Report—2010."
NOTE: All data are rough estimates.

KEY TAKE AWAY » Farmers markets primarily draw local residents.

market springs up a handful of blocks away. It's this "network" of farmers markets that may signal the death knell for traditional centralized food. Once this starts to occur, it will be harder and harder for traditional centralized food to "grow" volume, revenue, or profit in this geographic region. We suspect it is "easier" to kill reinvestment economics than nearly anyone might expect.

3. High price products

It's not just that farmers markets steal customers from the traditional centralized food network or that they steal a rapidly growing number of customers or that these thefts occur in very dense geographic regions. It's also that the customers are the highest-income customers who purchase the highest-profit products at grocery stores. We need to do more homework to get data that supports the point, but nearly everyone will concede that farmers market products displace very high price products from traditional supermarkets. More digging is necessary in this area, especially on the question, "Does the rise of high price volume on the distributed network increase cost for the centralized network by creating more bidding for very specific products?"

4. Customers are not very price sensitive

One of the more interesting trends I have observed in my research is that end-customers will not generally adopt a new product if the economics are bad (i.e., if the all-in price of the new product or new channel is above the all-in price of the old product or old channel). However, once the economics are attractive (i.e., when IRR from the new product or channel is above the discount rate of the customer), then the customer becomes largely price insensitive. We definitely see this in the case of farmers market customers. They say that "price" is not generally a key determinant of their purchase decisions. Take a look at the black bar in the next figure. What's interesting is that many people might ask, "How can a customer pay 50% more at a farmers market than at a supermarket and have a positive economic return?" This is such a great example of new networks replacing old networks. *Yes*, the price of a farmers market organic tomato is higher than the *average* tomato in the *average* supermarket, but it is often *less* than the price of a very high-end organic tomato in a high-end grocery store. So the customer is actually saving money. This is like someone purchasing a Prius. A Prius would be an expensive substitute for a compact Honda (to which many people might compare the Prius), but it is an inexpensive substitute for a mid-size BMW (to which many Prius buyers compare the Prius). **The key is to make sure that we are making the tradeoff analysis from the perspective of the buyer, not from the perspective of other people in the market.**

RANKING ATTRIBUTES BY FARMERS MARKET CUSTOMERS »
(% OF CUSTOMERS WHO HIGHLY RANK EACH ATTRIBUTE)

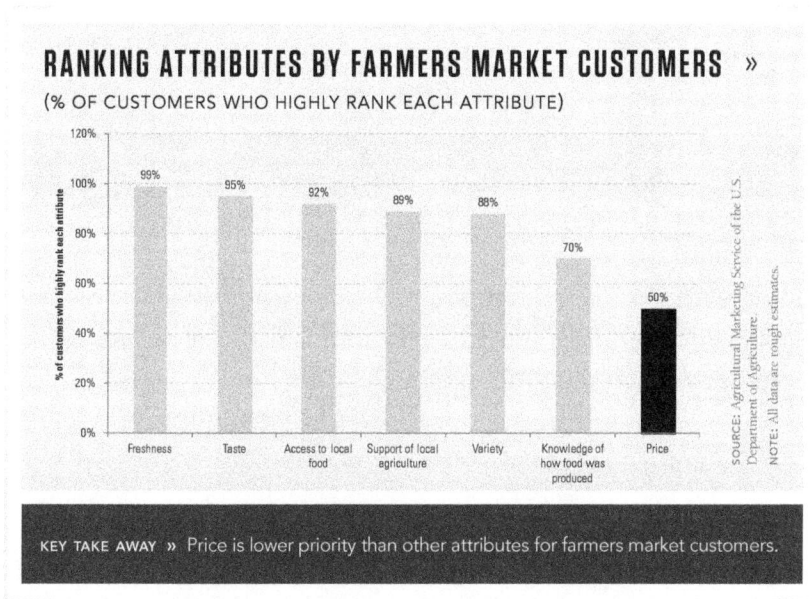

KEY TAKE AWAY » Price is lower priority than other attributes for farmers market customers.

One anecdote that suggests customers are not very price sensitive is that they appear to be purchasing more at farmers markets. Interviews with several farmers market managers suggest that average purchase per customer has risen significantly over the last 5 years. For example, the Union Square farmers market reports a sizeable shift in average purchase in the figure that follows.

CUSTOMER ASSESSMENT OF TOTAL PURCHASE[6] »

(% OF CUSTOMERS SPENDING PER CUSTOMER PER WEEK AS ESTIMATED BY CUSTOMER)

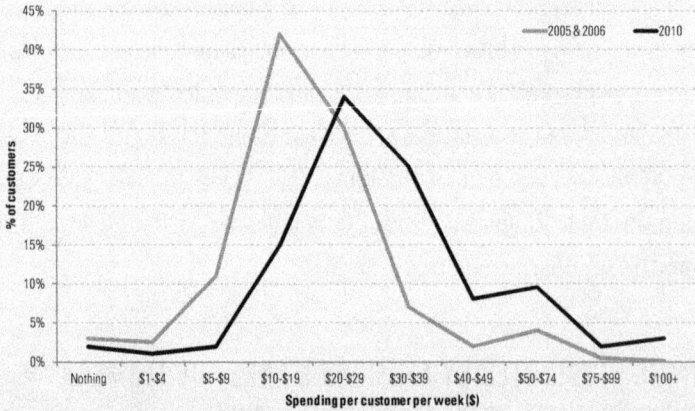

SOURCE: PHOTON Consulting, LLC, based on data from "Union Square Farmers Market Report—2010."

KEY TAKE AWAY » Data suggest increased spending per customer.

Data suggesting that these self-reported estimates may be accurate may be found by checking average weekly sales estimates from self-reported data ($31.27 per customer per week in 2010) compared to terminal data aggregated for credit card customers ($27.18 per customer per week in 2010 not including any cash purchases by those customers).

5. Trend is augmented by defections of existing customers across other similar products (i.e., there may be multiple negative network effects occurring at once)

Interestingly, the same customer that is substituting farmers market produce for supermarket produce may also be going through other similar substitutions. Three that come to my mind are:

○ **More substitution of self-grown food.** There has been a massive uptick in self-grow at rates that may be even faster than growth in farmers markets.

○ **More substitution of distributed water purification instead of purchasing water at the supermarket.** This is another trend with deployment characteristics similar to solar power but with impacts on supermarkets, convenience stores, bottled water distributors, and so on.

○ **More meals from food trucks.** This is another trend similar to farmers markets that will impact both restaurants and the prepared food sections of supermarkets.

Initially, Frank and I agreed when he wrote, "IF grocery chains lost 20% of the highest margin demand (and a key destination category for consumers) it would materially impact profitability and compel further in-store cost cutting. Is that plausible? In the short term: NO. I think it is a stretch to suggest this would happen anytime soon. If a grocery store is doing $6M, can a farm stand model siphon off $1.2 M in a local market within a 4 to 6 mile trading radius? I doubt it. Moreover, what likely really happens (I am guessing) is the total category sales grow as people are inclined to consume more rather than substitute. Within reason, there is an expandable consumption aspect to the category, aka, I eat more bananas when I have bananas."

However, after more thought, research and discussion, we have become increasingly convinced of the following: IF a very small number of supermarkets lose 20% of their operating profit to farmers markets, would the financial markets take notice? In the short run, we believe the answer is YES. Will this take down ALL of U.S. centralized food? NO. Might it impact MUCH more than just the supermarkets directly competing with farmers markets? YES.

The point of sharing Frank's background and our joint exploration of possible negative network effects in centralized food is to provide a clear example of how the right ingredients enable good "strategic marketing" insights to emerge. Without Frank's expertise (including his significant professional experience as an executive in centralized food and retail), personality and skills, I would not have been able to gain the insights we have developed about the possible implications of farmers markets (and other distributed food trends) for centralized food networks. Through this story and through our regular interactions, Frank demonstrates *all* of the ingredients on the long list at the start of the chapter:

1. **Flexible thinking:** Frank dives into new topics in each meeting with remarkable agility, moving from one mode of thinking about a topic to another.

2. **More interested in getting to the right answer instead of being right:** Frank regularly demonstrates that he is committed to finding the right answer instead of defending his answer as "right."

3. **Deep curiosity on a broad set of topics:** During our lunches and email exchanges, Frank and I regularly discuss energy, food, health care and a broad set of other industries and topics.

4. **Both big picture vision and attention to detail focus:** Frank will often jump from high-level conceptual thought to very specific data points then back in order to get different perspectives on a topic.

5. **Leverages other people's ideas:** Frank regularly asks for advice on business topics and then writes follow-up notes sharing how he has applied someone else's idea.

6. **Adjusts his ideas and admits initial answer is wrong if a better answer emerges:** The example in this chapter of Frank shifting from one way of thinking about negative network effects in centralized food

to another way of thinking is a good example that he admits when initial answers are wrong and adopts a superior answer.

7. **Defends his views fiercely in the absence of stronger logic:** The logic of negative network effects in the centralized food industry may not be correct, but it is the best hypothesis we have to date and, as a result, this is the view that Frank currently defends.

8. **Precise observation and detailed memory:** Frank can review a new topic and quickly identify and remember very specific details.

9. **Strong analytical aptitudes:** Per Frank's email to me with initial estimates of the revenue impact of farmers markets on grocery stores, he is able to rapidly execute mental math and deeper analysis.

10. **Tight reasoning capabilities:** Frank is highly logical and uses this logic to deduce answers that may not always be apparent from initial observations

11. **Inspirational written and oral communicators:** As leader of his company and as a discussion partner with me, Frank's communication skills are clear and his impact is motivational for others.

12. **Flexible interactions with diverse personalities:** I have observed Frank interact with a broad set of personality types with the acumen of a trained diplomat.

13. **Understanding of complex multinational business environments:** Frank has previously worked in several foreign countries and currently works closely with multinational teammates.

14. **Understanding of role of public policy in business:** As CEO of a health care education company, Frank is intimately involved in public policy debates that impact his business.

15. **Interested in impact slightly more than right answer:** Frank's fast and sustained upward career trajectory would have been impossible unless he cared deeply about impact.

16. **Progresses rapidly from "unaware" to "aware" to "understand" to "believe" to "act":** While some people might skip some of these steps in order to get to "act" faster, Frank instead goes through each step but does so with remarkable speed.

17. **Successful enough to have significant responsibility and true authority, but not so successful as to reduce drive:** Frank's professional accomplishments are impressive, but he still has not achieved a level of financial or professional success that is yet approaching even the bottom range of his aspirations.

18. **Unflinching commitment to succeed despite challenges:** Frank's drive to turn his company around during the recent financial crisis speaks to his commitment to succeed despite challenges.

19. **Deep personal integrity:** Frank is accurate, keeps his promises, and lives his personal and professional life by a high code of conduct.

20. **Puts needs and feelings of others ahead of his own:** Frank's manifold lessons to and mentoring sessions for me despite his overloaded schedule and heavy commitments speak volumes about how he places the interests of others before his own.

Overall, I am trying to signal to other CEOs that it is important to make sure you have the best ingredients in order to prepare a meal. There are significant challenges and opportunities on our collective horizon caused by negative network effects that require leadership, new modes of creating value and rededication of those with the most talent to inspire us all toward new achievements. Companies will need people like Frank to engage at a scale commensurate with their skills. In the coming 3 years, we face a broad set of difficulties that span economic, social, political and cultural arenas. These challenges will require many Franks alongside many other leaders to find and implement solutions.

Building a plan for the next 1,000 days that incorporates the risks and opportunities created by negative network effects requires CEOs to search for people with the right talents to help develop the plan and then execute it. In particular, it is important to pay attention to ingredients that enable "strategic marketing" insights because these are the ingredients that enable a company to most fully predict, prepare and benefit from explosive growth. If your company does not have the full suite of ingredients displayed by Frank, it is important to honestly assess your situation and commit to recruiting those resources. To paraphrase Theodore Roosevelt: If you want to enter the arena of negative network effects, it is mission critical that you come with the resources necessary to fight and win. You will need to avoid being cold or timid if you want to succeed in a world of negative network effects. Once you commit to ensuring you have the right ingredients and resources, then it is possible to truly address the questions, "Where is value?" and "How to capture it?"

Chapter 6

1,000 DAY PLAN

Don't be afraid to give up the good to go for the great.
—*John D. Rockefeller (1839–1937)*

1,000 DAY PLAN

Rockefeller, Vanderbilt, Carnegie and Gates were company leaders who were able to answer the questions, "Where is value?" and "How to capture it?" What's most interesting to me is that each of these business leaders answered these questions *before* their industry really existed. Their strategic marketing insights enabled them to look at economic and technology trends, qualitatively assess the potential emergence of an industry, quantitatively account for the growth of the industry, identify the areas of greatest value creation within that emerging industry and devise business models to capture value within that industry. For each, implementation was aggressive and highly profitable. In each case, there was at least one period of 1,000 days in which profit jumped exponentially and several 1,000 day periods in which the company sustained rapid profit growth.

NET PROFIT BY COMPANY »
($ MILLION)

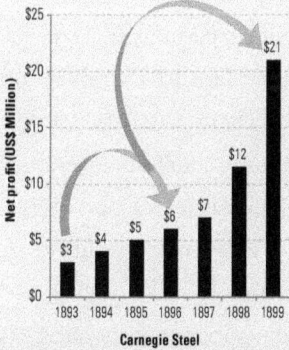

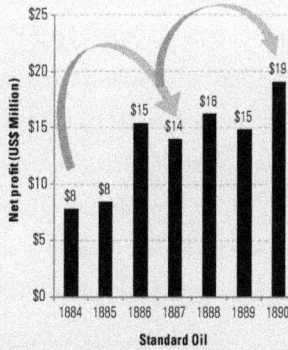

SOURCE: PHOTON Consulting, LLC, based on data from Charles Morris's *The Tycoons: How Andrew Carnegie, John D. Rockefeller, Jay Gould, and J.P. Morgan Invented the American Supereconomy* and Allen Nevins's *John D. Rockefeller.*

NOTE: All data are rough estimates

KEY TAKE AWAY » Rapid growth in net profit enabled by answering the questions "What industry is this?", "Where is value?", and "How to capture it?".

But before these "tycoons" shaped their industries (before Rockefeller shaped the global oil industry, Vanderbilt shaped the U.S. railroad industry, Carnegie shaped the global steel industry and Gates shaped the global software industry), each had a clear answer to the question, "What industry are we in?" This is a central question I want CEOs across a broad set of industries to ask themselves. My gut and analysis say that negative network effects are real, that their impacts will become apparent in a broad set of industries in the next 1,000 days and that there will be rapid fundamental transformation of many industries. This is a transformation of existing industries (especially centralized infrastructure industries) into "distributed infrastructure industries."

The "massive movement" discussed in Chapter 3 is already starting in areas where costs for distributed infrastructure are below the price of high-end products available from traditional centralized infrastructure networks. As discussed earlier, this is what we observed in First-Class mail converting to online bill pay and high-price electricity customers converting to rooftop solar power. The battle is between traditional centralized infrastructure (built on a foundation of centralized low-cost manufacturing leveraging economies of scale combined with low-cost distribution leveraging positive network effects that overcome inefficiencies in not perfectly matching production with use) and new distributed infrastructure (built on a foundation of customized-to-use production with a long-term commitment at the point-of-use).

This movement creates potentially huge business opportunities. These money making business opportunities and what they mean for CEOs across industries is a focus through the remainder of the book. For nearly a decade, I have been discussing negative network effects and the disruptive implications of distributive infrastructure with a broad set of CEOs/senior executives in the energy and chemicals industries and with a smaller set of CEOs/senior executives in the food, water, health care and communications sectors. When I recently sent an email sharing the themes discussed in the preceding paragraphs to one CEO, he wrote back:

MR,

I assume you saw the *WSJ* [*Wall Street Journal*] front page article this week about the rationalization of post offices, as you all predicted in the P-Magazine [PHOTON International magazine] piece last year. When I read the *WSJ* article, it felt like the introduction to a future award-winning *HBR* [*Harvard Business Review*] article that could be titled:

"The Emergence of the Distributed Infrastructure Industry"

. . . finally made possible at scale due to the changing economics across scores of industries from renewable energy industry to the grocery industry and even health care . . . This DII phenomenon is poised to not simply "dent" countless industries, but instead drive fundamental changes in the value chains of multi-billion dollar sectors of the global economy.

Forward thinking leaders are starting to make "game changing" business adjustments in anticipation of these new economics. For example, Sharp Solar, a leading energy provider founded by Tokuji Hayakawa, predicted over 50 years ago generating electricity in new ways "would benefit humankind to an extent we can scarcely imagine," is taking bold steps through acquisitions such as Recurrent Energy. Sharp aims to become a total solution company in the photovoltaic field, extending from developing and producing solar cells and modules to developing and marketing power generation plants.

Similarly, in the medical industry . . .

Good stuff, thanks for sharing; more to think about . . .

CEOs must do homework *before* diving into building their plan of attack. CEOs need to recruit talent (the "right" human resource ingredients as discussed in Chapter 5) and, like the work Frank and I have done in the food industry, need to evaluate *whether* negative network effects are likely to have an impact in their industry. To be clear: This *is* the CEO's job (not the job of someone else within the company) because most companies are not prepared for this undertaking. If a company is going down this path it will be led by the actions (no the orders) of the CEO. My conviction is that *most* CEOs in *most* existing industries who do their homework will find that negative network effects create large-scale potential opportunities and/or risks within the next 3 years. It will take time for CEOs to move from "unaware" to "aware" to "understand" to "believe" on the question, "Will negative network effects significantly impact my company?"

Once a CEO has done this homework and comes to "believe" that negative network effects may significantly impact the company, then the next step is to carefully answer the question, "Where is value?" Answering this question will require the CEO (perhaps with a small group of executives possessing the ingredients described in the previous chapter) to think like a "sector accountant." Basically, the CEO will need to put together a financial model for the sector that takes into account volumes, prices and costs in order to estimate revenue pools, profit pools and cash flows. While this type of accounting work might seem "beneath" a high-powered CEO, it is important to emphasize that accounting is at the core of answering the question, "Where is value?" As an example, throughout his career Rockefeller relied on bookkeeping and accounting skills (he was a bookkeeper and accountant for the early part of his career) to personally evaluate prices, volumes, revenue pools, profit pools and cash flows.

Implementing sector accounting is not rocket science. It requires significant work, but the work is often straightforward, especially for people possessing the 20 ingredients from Chapter 5. The goal of sector accounting is to put together financial statements (income statement, balance sheet, statement of cash flows) for the sector. This Chapter is a deep dive into the construction of a 1,000 day plan. This involves building the model for sector accounting. The starting point is the income statement, which is composed of five key rows:

1. Volume
2. Price
3. Revenue (i.e., volume X price)
4. Cost
5. Profit (i.e., revenue – cost)

Because revenue (3) and profit (5) result from the other key rows, the sector accountant really only needs to develop estimates for volume (1), price (2) and cost (4). This is quite straightforward. For a 1,000 day play, the CEO will need 1,000 days (~3 years) of history and 1,000 days (~3 years) of forecast. In the example table here, the key inputs are simply the three gray rows. Basically, the main focus for the CEO will be to quickly fill in the 21 gray cells with reasonable estimates and then to improve these estimates over time. The starting point for the 1,000 day plan should take no more than 10 days to complete.

BASIC INCOME STATEMENT FOR SECTOR ACCOUNTING »
(OVERVIEW)

	HISTORY OF 1,000 DAYS			NOW	FORECAST OF 1,000 DAYS			
	YEAR -3	YEAR -2	YEAR -1	YEAR 0	YEAR 1	YEAR 2	YEAR 3	
(1) VOLUME								
(2) PRICE								
(3) REVENUE								
(4) COST								
(5) PROFIT								

SOURCE: PHOTON Consulting, LLC.
NOTE: All data are rough estimates.

KEY TAKE AWAY » Starting point: Volume, price and cost.

To develop estimates for these 21 gray cells, the CEO-as-sector-accountant starts by building a supply curve (A) and then a demand curve (B). As a starting point, play a mental exercise without any research: Draw a best-guess-without-any-information supply curve and demand curve for the current year (Year 0). While this might be a "mission impossible" request for many people, a senior business executive should be able to make a quick estimate, though this may require some of the "explosive ingredients" mentioned in the last chapter (i.e., combination of flexible thinking, deep curiosity, strong analytical aptitudes and tight reasoning capabilities). Then, again without any research, draw best-guess supply and demand curves for 1,000 days (Year −3) prior and 1,000 days (Year 3) later. These six curves (supply and demand curves for Year −3, Year 0 and Year 3) should be drafted within 1 day, preferably within 1 hour. This is a "fast-draw" exercise without any research. Welcome to the Wild West: If you don't draw fast, then you die or are robbed.

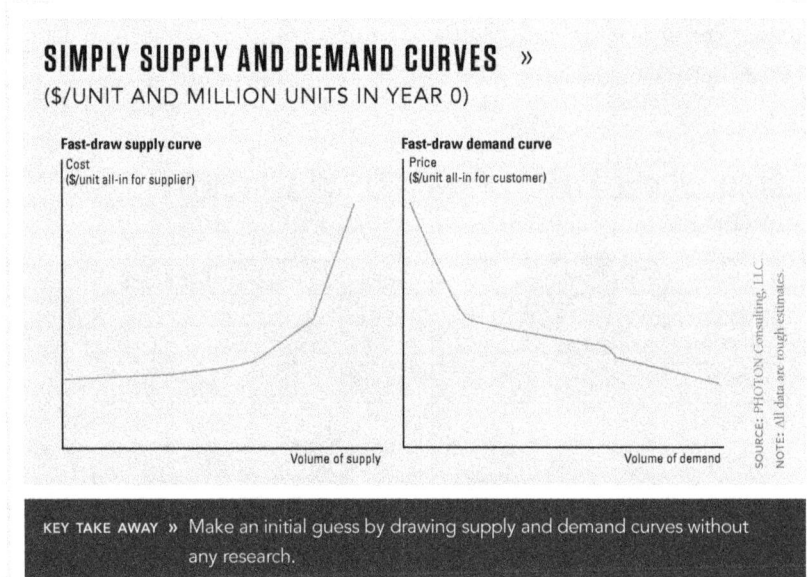

SIMPLY SUPPLY AND DEMAND CURVES »
($/UNIT AND MILLION UNITS IN YEAR 0)

Fast-draw supply curve
Cost
($/unit all-in for supplier)

Fast-draw demand curve
Price
($/unit all-in for customer)

Volume of supply

Volume of demand

SOURCE: PHOTON Consulting, LLC.
NOTE: All data are rough estimates.

KEY TAKE AWAY » Make an initial guess by drawing supply and demand curves without any research.

With these fast-draw curves as a mental background, the next step is to dive into more detail on the supply curve (A). To estimate the supply curve, you will want to collect (in a spreadsheet) a list of companies (a1) with their production volume (a2) for either Year 0 or Year 1. In addition, you might collect capacity (a3), but this is less important until later. This list can normally be compiled within 1 week (preferably within a couple days) using the Internet. The goal is *not* a 100% complete list of companies or production or capacity, but instead the best list that can be compiled quickly. The goal of compiling this list is to then be able to have some modest detail that enables the CEO to be "aware" of production volume by company in order to be incrementally more informed about the supply curve's volume of supply (A; x-axis). Typically, estimates for 5 to 20 companies are sufficient to begin getting a sense of scale for the supply curve's x-axis (A) in Year –1 or Year 0. The next figure offers an example of production volume by company. It is important to emphasize that this data collection is a CEO activity. Because the industry is not yet well defined, leaving this initial data in the hands of a more junior team member is like letting an 8-year-old drive your car: You might not have an accident, but it's an unnecessary risk for something that doesn't take much effort. Basic message to CEOs: Get your hands dirty. This is the starting point to reshape the direction of your company and your industry. Don't trust it to someone else. Spend 10 days (or less) on this.

PRODUCTION VOLUME (EXAMPLE: 2010 SOLAR POWER MODULE PRODUCTION) »
(MILLION WATTS/YEAR)

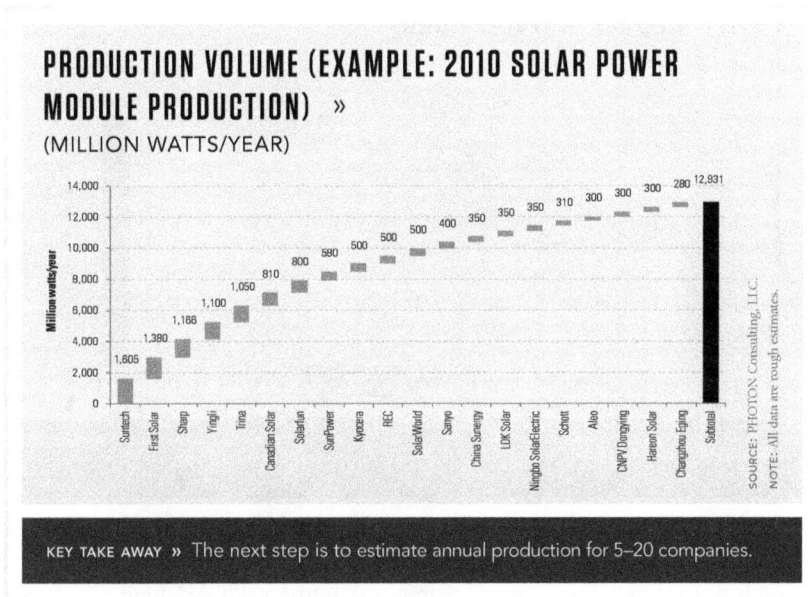

KEY TAKE AWAY » The next step is to estimate annual production for 5–20 companies.

Then do a similar exercise for the demand curve (B). To estimate the demand curve, you will want to collect (in a spreadsheet) a list of markets (b1) with their end-customer purchase volume (b2) in either Year 0 or Year 1. In addition, you might collect similar information on specific segments within geographic markets (b3) if/when easily available, but this is less important until later. This list also can normally be compiled within 1 week (preferably within a couple days) using the Internet because the goal is the best list that can be compiled quickly. The goal of compiling this list is to then be able to have some modest detail that enables the CEO to be "aware" of customer purchase volume by market in order to be incrementally more informed about the demand curve's volume of demand (B; x-axis). Typically, estimates for 5 to 20 markets are sufficient to begin getting a sense-of-scale for the demand curve's x-axis (B) in Year −1 or Year 0. The figure that follows shows an example of customer purchase volume by market.

END-CUSTOMER PURCHASE VOLUME (EXAMPLE: 2010 SOLAR POWER MODULE INSTALLATIONS) »
(MILLION WATTS/YEAR)

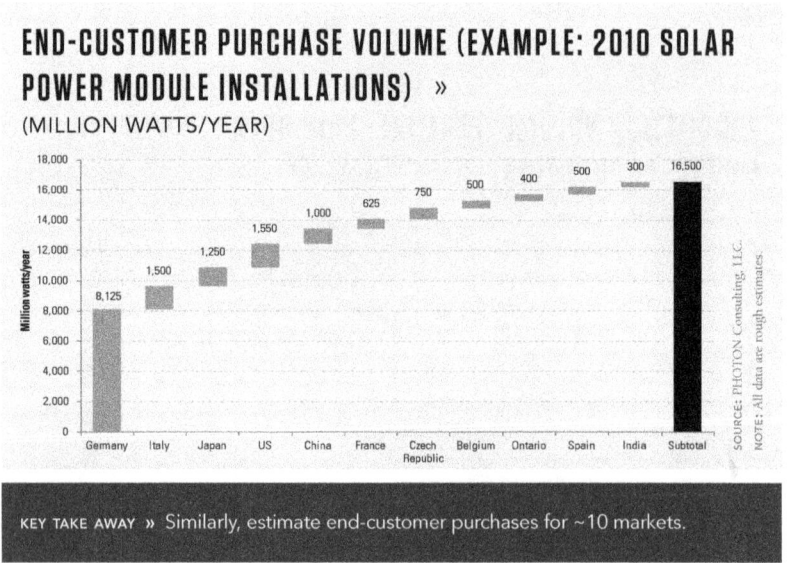

KEY TAKE AWAY » Similarly, estimate end-customer purchases for ~10 markets.

With this information in mind, then go back to quickly (<1 hour) revise the six initial fast-draw supply and demand curves. In this case, most of the revision to the fast-draw supply and demand curves may be to the scale of the x-axis, not yet to the scale of the y-axis or the slope of the curves. (Note: These revisions will require other "explosive ingredients" such as lack of ego, being more interested in getting to the right answer than being right and a willingness to adjust ideas if better answers emerge.) At this point, it is already possible to insert rough estimates for row volume (1) of the "Basic income statement for sector accounting." Start with the estimate of volume (1) in Year −1 or Year 0 based on the data in your supply (A) spreadsheet, and then also take a stab without deep research at the historical and future volume. With this, 7 of the original gray cells are drafted and only 14 remain.

The next step is to begin addressing price (2). This is done from two sides. The first side is from the perspective of suppliers. Collect estimates of prices by supplier (a4) in the supply (A) spreadsheet. You can do this by looking up price offers on the Internet (there are often price lists and price offers available), reviewing the results of publicly listed companies for disclosure about price, or calling suppliers to ask for rough/benchmark price quotes. When available, this information should be recorded for each company. When not easily available, this information should be skipped (for now). Normally, these prices will be for either Year −1 or Year 0. The second side is from the perspective of customers. Collect estimates of prices by end-market (b4) in the demand (B) spreadsheet. You can do this by looking up government statistics that track end-customer prices or by contacting end-customers that have purchased. When available, the information for Year −1 or Year 0 should be recorded for each market and, when not available, the information should be skipped. In addition, you can also update and expand information for (a1), (a2), (a3), (b1), (b2) and (b3) as you proceed with collecting estimates for (a4) and (b4).

With this information in mind, then go back to quickly (<1 hour) revise the six fast-draw supply and demand curves. In this case, most of the revision to the fast-draw supply and demand curves may be to the scale of the y-axis. In addition, it is now possible to begin drawing the demand curves (B) more precisely for Year −1 or Year 0 using "boxes." The following figure provides an example of the demand curve (B) drawn using the list of markets (b1) with the volume of end-customer purchases (b2) in each market and the price (b4) in each market. This provides the label for each box (b1), the x-axis for each box (b2) and the y-axis for each box (b3). In addition, the weighted average price of this "box" demand curve (i.e., multiply volume in each market times price in each market to get the revenue pool in each market, then add these market revenue pools to get an estimate of the global revenue pool and divide the global revenue pool by the global purchase volume in order to get a global weighted average price) can now be inserted as a rough estimate of Year −1 or Year 0 in row price (2) of the "Basic income statement for sector accounting." Start with the estimate of price (2) in Year −1 or Year 0 based on the data in your demand (B) spreadsheet and then also take a stab without deep research at the historical and future price. With this, 14 of the original gray cells are drafted with only 7 remaining.

"BOX" DEMAND CURVE (EXAMPLE: 2010 SOLAR POWER DEMAND CURVE) »

($/WATT END-CUSTOMER PURCHASES MILLION WATTS/YEAR)

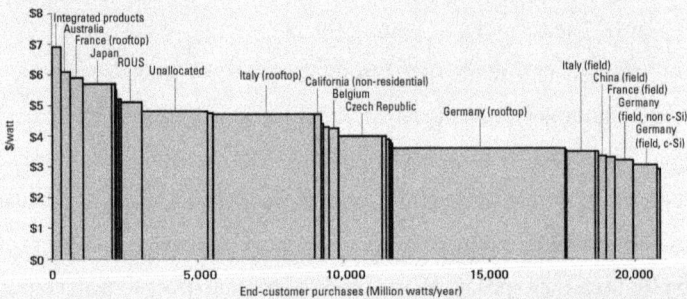

KEY TAKE AWAY » Use the data you now have to draw "box" demand curves.

The next step is to begin estimating cost (3). This is done from two sides. The first side is from the perspective of suppliers. Collect estimates of costs by supplier (a5) in the supply (A) spreadsheet. This can be done by reviewing the results of publicly listed companies for disclosure about cost or calling suppliers to ask for rough/benchmark cost estimates. (Note: Disclosures from pure-play companies focused only on this business are often most helpful, but larger companies also often disclose significant details about volumes, prices and costs.) As a starting

point, it is often helpful to define "cost" as "cost of goods sold (COGS) plus operating expenses (SGA)" that lead to earnings before interest and tax (EBIT) in common income statement accounting. In the case of cost by supplier (a5), we are
interested in the cost per unit for each supplier. This is equal to COGS plus SGA
divided by production volume. (Note: Later we will pay more attention to production volume versus shipping volume, but for now either production or shipping
volume is fine to use. Similarly, later we will pay more attention to more refined
breakdowns of cost.) In addition, academics, government agencies, news organi-
zations, financial market analysts and consultants often publish estimates of cost.
When available, this information should be recorded for each company. When not
easily available, this information should be skipped (for now). Normally, these
costs will be for either Year −1 or Year 0. The second side is from the perspective
of customers. The existing estimates of prices by end-market (b4) in the demand
(B) spreadsheet provide a baseline to estimate cost by market (b5). You can do
this by assuming a profit margin that is then subtracted from the end-customer
price in order to estimate cost by market (b5). This profit margin is typically 5 to
10% of end-customer price, but can vary significantly from industry to industry
and market to market. The point of estimating cost by market (b5) at this point
is only to provide a reality check to ensure that cost by company (a5) appears in
a reasonable range. When available, the information for Year −1 or Year 0 should
be recorded for each market and, when not available, the information should be
skipped. In addition, you can also update and expand information for (a1), (a2),
(a3), (a4), (b1), (b2), (b3) and (b4) as you proceed with collecting estimates for
(a5) and (b5).

    With this information in mind, then go back to quickly (<1 hour) revise the
six fast-draw supply and demand curves. In this case, most of the revision to the
fast-draw supply and demand curves may be to the scale of the y-axis. In addition,
it is now possible to begin drawing the supply curves (A) more precisely for Year
−1 and possibly Year 0 using "boxes." The following figure provides an example
of the supply curve (A) drawn using the list of companies (a1) with the volume
of production (a2) by each company and the cost per unit (a5) in each company.

This provides the label for each box (a1), the x-axis for each box (a2) and the y-axis for each box (a5). In addition, the weighted average cost of this "box" supply curve (i.e., multiply volume of each company times cost per unit of each company to get an estimate of total cost before interest and taxes for each company, then add the total costs for all companies and divide by the total production of all companies) can now be inserted as a rough estimate of Year –1 or Year 0 in row (4) Cost of the "Basic income statement for sector accounting." Start with the estimate of cost (4) in Year –1 or Year 0 based on the data in your supply (A) spreadsheet, and then also take a stab without deep research at the historical and future cost. With this, all 21 of the black cells are drafted.

"BOX" SUPPLY CURVE »

(EXAMPLE: 2010 SOLAR POWER SUPPLY—MILLION WATTS PRODUCTION AND $/WATT ALL IN COST)

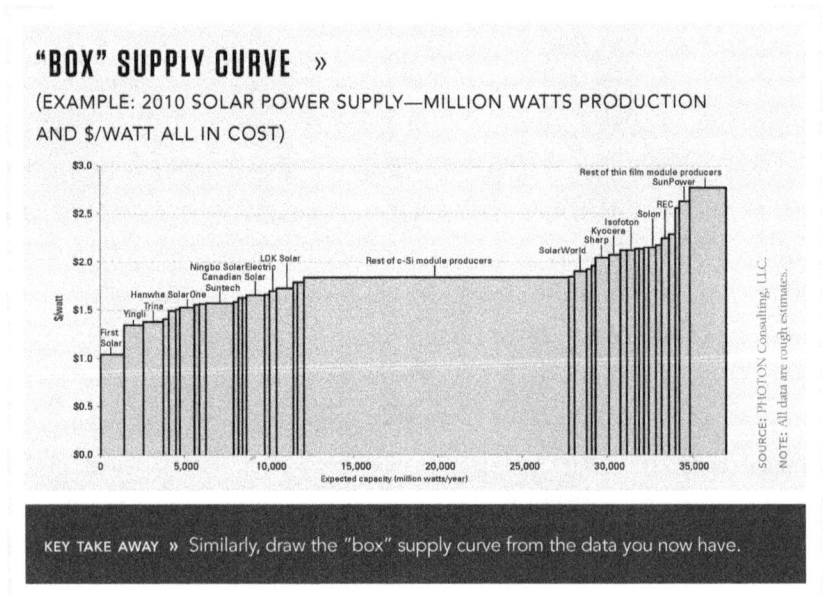

KEY TAKE AWAY » Similarly, draw the "box" supply curve from the data you now have.

By completing the formulas for revenue (3; volume X price) and profit (5; revenue – cost), the sector income statement is drafted at a high level and CEOs can congratulate themselves for having quickly (in 10 days or less) filled in a rough sector income statement. There are now "anchor" numbers (including a very rough estimate of revenue and profit pools) to center thinking and discussion about the sector. However, there are many, many splinters that require sanding. This is where the "explosive ingredients" of curiosity and unflinching commitment to succeed despite challenges become important. There is a lot to sandpaper, including:

○ Creating company-by-company estimates of volume, price and cost for the last 3 years, the current year and the next 3 years;

○ Creating market-by-market estimates of volume, price and cost for the last 3 years, the current year and the next 3 years;

○ Refining the company-by-company estimates of volume, price and cost to take into account each step of the supply chain;

○ Refining the market-by-market estimates of volume, price and cost to take into account segment-by-segment estimates of volume, price and cost;

○ Refining estimates of cost to take into account more granular subcategories of cost and the trends in these subcategories of cost for the last 3 years, the current year and the next 3 years;

○ Drawing and redrawing the supply and demand curves in both "fast-draw" and "box" form for Year –3, Year –2, Year –1, Year 0, Year 1, Year 2 and Year 3;

○ Refining volume estimates to take into account production and shipments at each step of the supply chain leading up to end-customer purchases;

○ Balancing these estimates of production, shipments and end-customer purchase volumes with estimates of inventory at each step of the supply chain;

○ Augmenting volume estimates of production, shipment and end-customer purchases with additional analysis of capacity and capacity additions;

- Extending from only sector income statement to also estimate sector balance sheet and statement of cash flow;

- Expanding from annual estimates to quarter estimates; and

- More carefully incorporating moves of currency exchange rates;

- More thoroughly normalizing accounting standards from diverse inputs (e.g., Japanese accounting standards and timing versus U.S. accounting standards and timing);

- More carefully tracking and estimating the impact of policies and potential changes in policies;

- Articulating answers to "What sets volume?" (i.e., volume-setting mechanisms) and "What sets price?" (i.e., price-setting mechanism) in each end-customer market and segment in each quarter and year;

- Establishing peer review processes;

- Developing "what if?" scenarios that adjust sector volume, price, revenue pool, cost and profit pool estimates based on different input assumptions about the macro-environment, substitutes and other factors; and

- Perpetually updating the sector accounting based on the latest information as time passes; and

- Perpetually evaluating the accuracy of sector accounting (especially forecasts) against the actual performance of the sector as time passes and incorporating learning into the sector accounting.

This effort will take ~100 days and is far too extensive and expensive for a CEO to undertake from the start. Instead, the CEO (like all good executives who are forced to make decisions without full information and know that they will have to change in the future) must drive toward improving their understanding quickly but incrementally. The CEO should continuously adjust and improve the high-level sector accounting discussed prior to these bullets, but should not misplace the forest for all the trees. The key question is, "Where is value?" When the CEO says, "I have a good sense of the answer to the question 'Where is value?'" then that CEO is probably ready to move on.

To get to this level of conviction, most CEOs want to keep driving toward deeper information. In my experience, there are three specific output images that are most useful for a CEO who wants more information about an industry experiencing explosive growth: (1) Price and volume adoption patterns at each step along the supply chain starting with the end-customer level, (2) maps of profit pools and (3) estimates of future cost reduction trends along the supply chain.

The first output image that may provide significant insight to the CEO is (1) price and volume adoption patterns at each step along the supply chain starting at the end-customer level. The next figure shows the adoption pattern of solar power in Japan from 1994 to 2010. The x-axis displays the price of solar power systems to end-customers in Japan after taking into account all prices and incentives. This is the "net price" facing the customer. They y-axis is the volume of new solar power systems adopted in Japan each year 1994 to 2010. The basic "story" of this figure starts in 1994 when the Japanese government launched the "New Sunshine Program" to provide end-customers with financial incentives to purchase rooftop solar power systems. In 1994, the net price of these systems was very high (roughly $0.70/kWh) and very few systems were installed. With government incentives, the net price of systems dropped and the volume of installations increased (albeit from very low volumes that are hard even to see on this figure). This trend of significant price decrease with some volume increase occurred from 1994 to 1996. Demand was elastic but not remarkably so. Then something changed in 1997. The elasticity of demand increased, meaning that a decrease in price drove a much larger increase in installations. This continued from 1997 to 2003, with price decreasing very little while volume jumped rapidly.[1]

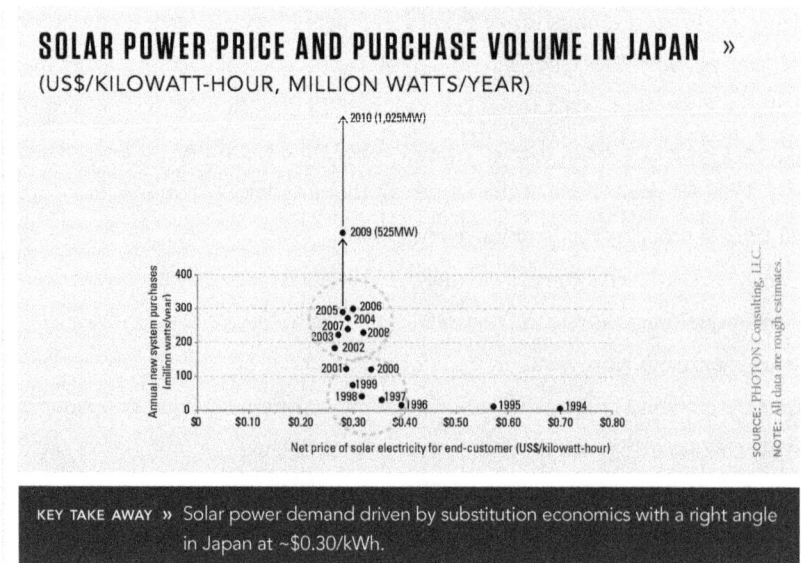

SOLAR POWER PRICE AND PURCHASE VOLUME IN JAPAN »
(US$/KILOWATT-HOUR, MILLION WATTS/YEAR)

KEY TAKE AWAY » Solar power demand driven by substitution economics with a right angle in Japan at ~$0.30/kWh.

This occurred with a price of solar power of around $0.30/kWh, which was roughly the average price of grid-based electricity in Japan. This is important and CEOs across the solar power sector have paid particular attention to this dynamic: A key mechanism for PV pricing is parity with grid electricity. When the price of solar power approaches the price of the traditional electricity grid, demand expands rapidly with only modest price decreases. (Note: This dynamic of parity with the price of the existing network is important for evaluating price-volume in other distributed infrastructure industries, too.)

The dynamic of 2004 is also important to CEOs across the solar power sector. With the introduction of solar power incentives in Germany (2004) then Spain (2005), demand for solar power installations surged in Europe. This surge in demand was at a price point in Europe above the price point in Japan. The result was that solar power modules were drawn to Germany and Spain, leaving installations in Japan to stagnate. If you look closely at the figure, Japanese solar power installations in 2008 were actually lower than in 2004. This is important and CEOs across the solar power sector have also paid particular attention to this dynamic: Demand in one geographic market has potential to significantly impact other geographic markets.

It's interesting to note that the rapid growth of Spain's solar power market led to a change of Spanish policy that significantly reduced solar power installations after 2008. This was discussed in Chapter 3. The result in Japan is stunning. With modules no longer "attacking" Spain, there was available supply for Japan and new Japanese installations jumped more than 2X in 2009 and nearly doubled again in 2010.[2] From the perspective of executives in the solar power sector, the graphical representation of this volume/price dynamic is very helpful because it enables CEO conviction that installation volume growth and prices will be very strong as the price of solar power nears the price of grid electricity. With this conviction, the CEOs "believe" that there is sizable value to be captured and they begin looking more carefully to understand the dynamics of, "What sets price?" and, "What sets volume?" in specific markets. For them, this is *not* an academic exercise. They are looking to this analysis as a trigger to move quickly from "aware" to "understand" so that they can then move to "believe" and "act."

For CEOs outside the solar power sector who are evaluating "Where is the value?" in an industry that may be impacted by negative network effects, similar analysis of the price-volume tradeoff (graphically presenting volume and price by market while asking the questions, "What sets price?" and "What sets volume?") will go a long way to moving them toward "understand" and then "believe" in the potential profit pool awaiting them. A graphical representation of the volume-price dynamic also goes a long way to emphasize the importance of "strategic marketing" because understanding the demand dynamics across many markets is necessary to evaluate the demand dynamics within a specific market. My personal observations suggest that this **highly elastic demand dynamic is a common characteristic on the positive side of negative network effects and is the main driver of explosive growth for distributed infrastructure sectors. However, the companies that truly thrive**

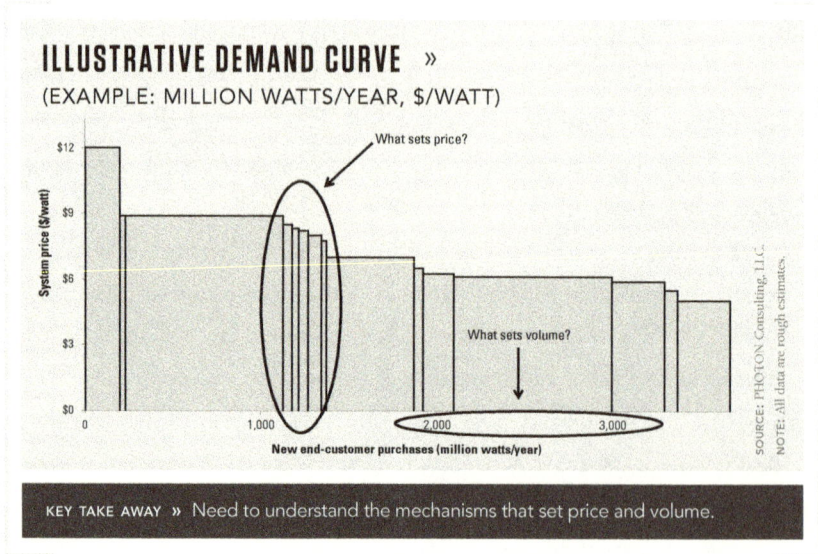

ILLUSTRATIVE DEMAND CURVE »
(EXAMPLE: MILLION WATTS/YEAR, $/WATT)

KEY TAKE AWAY » Need to understand the mechanisms that set price and volume.

<u>in an environment of negative network effects are those that truly understand</u>
<u>the global nature of the demand curve with interplay across markets.</u>

Yet even if a CEO comes to "understand" that there are big profit pools in a distributed infrastructure sector, the CEO needs more focused information in order to "believe" that those profit pools are accessible for a specific company. This is the second output image described earlier. Creating this output image requires more precise breakdowns of the profit pool. These breakdowns include:

- Profit pool by company
- Profit pool by geographic market
- Profit pool by customer segment
- Profit pool by technology
- Profit pool by step of the supply chain

Among these, profit pool by step of the supply chain is often the image that is most compelling to move CEOs from "understand" to "believe" a specific answer to the question, "Where is value?" What follows is one example of the profit pool by step of the supply chain from the solar power industry. A similar representation is to show operating profit margin (operating profit ÷ revenue in % terms). These specific representations can be compelling for a CEO because it is possible for the CEO to quickly check the profit pool (or the profit margin) at one step of the chain versus another step of the chain against the profit results of publicly listed companies participating at the different steps. Basically, the CEO can see the "top-down" estimate of the profit pool and compare it with a "bottom-up" company level estimate. Once the CEO establishes the connection between sector profit pool estimates and company profit results, the immediate question the CEO often tries to answer is, "How will the profit pool evolve in the coming years?" The CEO coming to "believe" in any answer to this question requires "understanding" the top-down sector accounting discussed earlier in the chapter combined with "understanding" the price-volume dynamics along the supply chain and "understanding" a thoughtful estimate of profit pool that is broken down into more precise buckets (e.g., by step of chain or by technology or by geography).

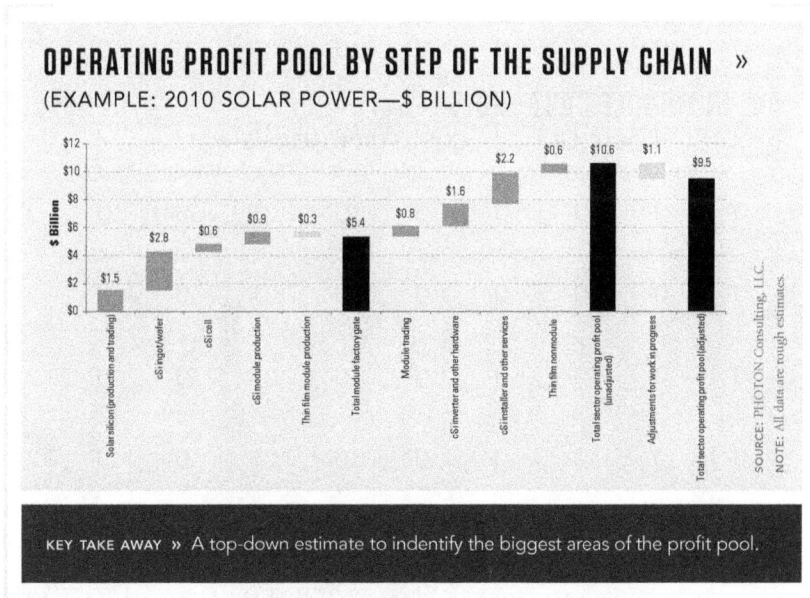

OPERATING PROFIT POOL BY STEP OF THE SUPPLY CHAIN »
(EXAMPLE: 2010 SOLAR POWER—$ BILLION)

KEY TAKE AWAY » A top-down estimate to indentify the biggest areas of the profit pool.

Often a CEO will search for the answer to one additional question before moving fully to "believe." This question is, "What are expectations for cost in the next few years?" There are two underpinnings of this question. On one hand, the CEO is concerned that cost reductions for the sector might slow, making it more difficult to execute on the dynamic of slightly lower price opening a lot of demand. On the other hand, the CEO wants to understand how difficult it will be for a specific company to reduce costs. Will a small subset of companies stand out on cost reductions or will nearly all companies be able to achieve significant cost reductions? In this situation, CEOs often look to cost benchmarking and evaluating cost best practices versus common practices. The next example presents an output image that may be useful. This analysis compares actual company cost results (estimates for publicly listed companies) with cost benchmarks of best practice cost at each step of the supply chain ("sum of bests"). The comparison highlights that even the lowest cost players ($1.31/watt) are more than 20% above actual cost performance ($1.08/watt) of best practice players along the supply chain. The implication is that there is likely significant room for cost reductions in the coming years, which may give the CEO added conviction in the resilience of the profit pool.

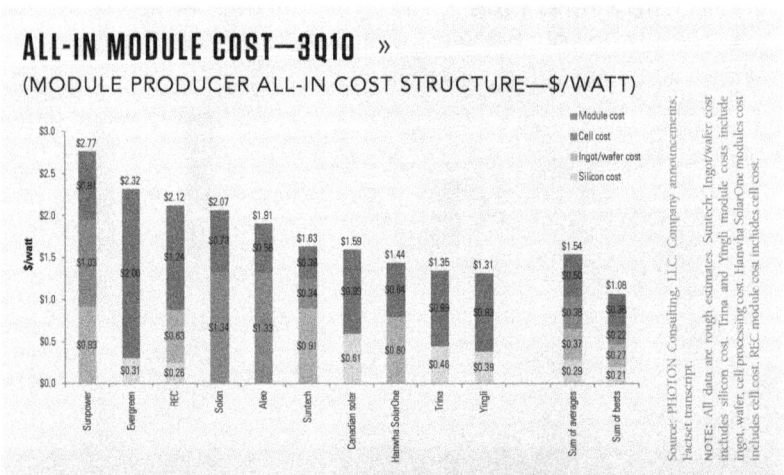

ALL-IN MODULE COST—3Q10 »

(MODULE PRODUCER ALL-IN COST STRUCTURE—$/WATT)

KEY TAKE AWAY » Cost benchmarking gives additional confidence in the resilience of the profit pool.

At this point, the CEO's answer to the question, "Where is value?" is increasingly clear. But the CEO should perpetually check this view against the views of others within the company and outside the company. This is when the "explosive ingredients" of being willing to leverage other people's ideas and having more interest in getting to the right answer instead of being right are of high importance.

This is also the point at which it is mission critical that being more interested in impact than in being right comes into play. While the assessment of, "Where is value?" may change, it is now time to quickly develop an answer for, "How to capture it?" This is the 1,000 day plan by which the CEO will drive the company from his sector-level assessment into company-level action. Developing this company-level action plan (1) starts with assessing the sector value map in comparison with different business models the company might pursue; (2) then compares these potential business models for fit with the company's skills, capabilities and resources; (3) then prioritizes a potential business model for deeper assessment; and (4) finally builds a 1,000 day action map for each function of the company that ties back to the core drivers of value capture.

As a first step, the CEO should (1) assess under different scenarios the different business models the company might pursue. The dynamic nature of the distributed infrastructure sectors means that CEOs must be prepared to quickly and decisively adjust their "bets" in response to industry changes. In particular, executives must be prepared to quickly adapt when a policy, financial market, or technology change significantly alters the direction of the industry or the distribution of value-capture, risk, or opportunity across the supply chain. From our standpoint, one of the key first steps for achieving this preparation is developing sector-level and company-level scenarios based on different assumptions about the most important factors likely to drive value in your sector. By defining the game and quantifying the potential risks and rewards in a given scenario, decision makers are in a much better position to place informed, well-calculated bets about the best direction for their business.

The CEO needs to develop a set of quantified scenarios for how the sector might evolve. Doing this requires identifying the key uncertainties facing the sector. In the solar power sector, for example, key uncertainties include policy, capital availability, technology evolution, exchange rates and the price of the substitute (grid electricity), among other uncertainties. The CEO should carefully identify key uncertainties in the sector and carefully think through how different plausible developments of these uncertainties could unfold over the next 1,000 days. The goal is to have each scenario represent an internally consistent story for how the sector might evolve based on plausible assumptions for each of the uncertainties. This is an example of scenarios developed for the solar power sector.

KEY UNCERTAINTIES BY SCENARIO »
(EXAMPLE: SCENARIO ASSUMPTIONS FOR SOLAR POWER SECTOR)

	"Current hand"	"Policy flushed"	"Policy all-in"	"Financial fold 'em"	"Thin wins"
Policy	Controlled expansion	Key growth markets capped	Significant global expansion of PV policy support	Controlled expansion	Controlled expansion
Capital availability	No change in interest rates from current levels. Growing capital availability for installation financing and capacity growth	No change in interest rates from current levels. Growing capital availability for installation financing and capacity growth	No change in interest rates from current levels. Growing capital availability for installation financing and capacity growth	400bps end-customer risk premium increase in 2011–2012, interest rates normalize in 2013–2014. Significant capital constraints for installations and capacity growth in 2011–2012	No change in interest rates from current levels. Growing capital availability for installation financing and capacity growth
Technology evolution	No major advancement/disruptive improvement in 'PV complementary' or non-PV technologies	No major advancement/disruptive improvement in 'PV complementary' or non-PV technologies	No major advancement/disruptive improvement in 'PV complementary' or non-PV technologies	No major advancement/disruptive improvement in 'PV complementary' or non-PV technologies	Revolutionary TF advancement. No major advancement in 'PV complementary' technologies
Exchange rate assumptions	Same across all cases	Same across all cases	Same across all cases	Same across all cases	Same across all cases
Grid price assumptions	2.5–3.0% increase p.a.	2.5–3.0% increase p.a.	2.5–3.0% increase p.a.	2.5–3.0% increase p.a.	2.5–3.0% increase p.a.

SOURCE: PHOTON Consulting, LLC.
NOTE: All data are rough estimates.

KEY TAKE AWAY » Scenarios are internally consistent stories for how the sector might evolve.

Each scenario should then be quantified to evaluate "Where is value?" within the sector. This quantification includes volume, price, revenue, cost and profit estimates for each step of the supply chain. The goal of quantification is to enable comparison of the sector's profit pool (including subsets of the sector's profit pool) under various scenarios. Here is an example of the operating profit margin estimates for solar power manufacturing ("GWA manufacturing through FG OPM %"), Chinese Tier 1 module manufacturers ("CT1 manufacturing OPM %"), non-module suppliers (Non-module OPM %) and the overall sector including the entire supply chain through customer purchase (Sector level OPM %). The usefulness of this tool is to identify major areas of risks (e.g., significant negative operating profit in 2011–2012 for average solar power manufacturers under assumptions of the "Financial fold 'em" scenario) and opportunities (e.g., significant positive operating profit for Chinese Tier 1 manufacturers under assumptions of "Policy all-in" scenario) that hinge on how specific uncertainties evolve.

OPERATING PROFIT MARGIN BY SCENARIO »
(EXAMPLE: OPM % BY SOLAR POWER SECTOR SCENARIO)

GWA manufacturing through FG OPM (%)	2008	2009	2010	2011	2012	2013	2014
"Current hand"	43%	16%	7%	5%	4%	6%	5%
"Policy flushed"	43%	16%	7%	3%	-8%	0%	0%
"Policy all-in"	43%	16%	7%	22%	29%	31%	34%
"Financial fold 'em"	43%	16%	7%	-21%	-17%	-3%	6%
"Thin wins"	43%	16%	7%	5%	10%	3%	14%
CT1 manufacturing OPM (%)	**2008**	**2009**	**2010**	**2011**	**2012**	**2013**	**2014**
"Current hand"	0%	21%	17%	17%	15%	15%	14%
"Policy flushed"	0%	21%	17%	16%	2%	10%	7%
"Policy all-in"	0%	21%	17%	25%	28%	30%	33%
"Financial fold 'em"	0%	21%	17%	-11%	-5%	1%	9%
"Thin wins"	0%	21%	17%	17%	13%	5%	-5%
Non-module OPM (%)	**2008**	**2009**	**2010**	**2011**	**2012**	**2013**	**2014**
"Current hand"	22%	14%	15%	16%	19%	19%	19%
"Policy flushed"	22%	14%	15%	11%	23%	24%	21%
"Policy all-in"	22%	14%	15%	9%	12%	17%	23%
"Financial fold 'em"	22%	14%	15%	18%	10%	28%	26%
"Thin wins"	22%	14%	15%	16%	9%	10%	7%
Sector level OPM	**2008**	**2009**	**2010**	**2011**	**2012**	**2013**	**2014**
"Current hand"	34%	15%	11%	11%	13%	14%	14%
"Policy flushed"	34%	15%	11%	7%	13%	16%	13%
"Policy all-in"	34%	15%	11%	15%	20%	24%	28%
"Financial fold 'em"	34%	15%	11%	3%	-1%	17%	18%
"Thin wins"	34%	15%	11%	11%	10%	7%	10%

SOURCE: PHOTON Consulting, LLC.
NOTE: All data are rough estimates.

KEY TAKE AWAY » Scenarios are useful to identify areas of major risk and opportunity.

The scenarios should then be applied to specific business models. The goal is to qualitatively and quantitatively evaluate how different business models might deliver or destroy value under various scenarios for how the sector might evolve. These potential business models are different rationales for how the company might pursue and capture value. In general, we have observed three "prototypes" of business models that create significant value in distributed infrastructure industries:

◦ **Low-cost manufacturing business models:** These are business models that enable a company to be at or near the left side of the supply curve.

◦ **Differentiated service business models:** These are business models that enable a company to create distinctive relationships with customers that drive lower customer defection rates and/or enable higher prices.

◦ **Differentiated transaction/action business models:** These are business models that enable a company to more quickly move in and out of actions and asset positions.

In this next example, the business models include polysilicon/wafer manufacturing, Chinese traditional module manufacturing, European traditional module manufacturing, new company new technology module manufacturing, providing equipment (i.e., supplying manufacturing equipment to manufacturers), installing solar power systems in Europe and installing solar power systems in the U.S.

OPERATING PROFIT BY SCENARIO »
(EXAMPLE: OP ($ MILLION) BY SOLAR POWER SECTOR SCENARIO)

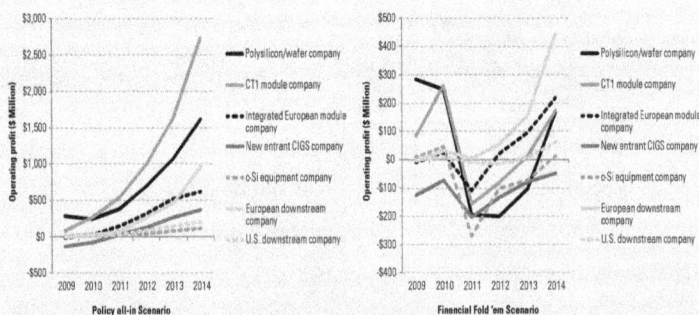

KEY TAKE AWAY » Evaluate performance of different business models under different sector scenarios.

Under the scenario assumptions presented earlier, the two scenarios presented (i.e., "Policy all-in" and "Financial fold 'em") result in very different value capture for the different business models. By evaluating the performance of different business models under different sector scenarios, the CEO is able to get a much better sense for the opportunities and risks of pursuing specific business models.

The CEO should then assess the fit of the various business models with the skills, capabilities and resources of the company. Whereas there are *many* business models that a CEO may *want* to pursue, there are only some that the company has the capabilities/skills to pursue, and then only a subset of these does the company have the resources to pursue. Given the massive changes occurring within the industry, it is likely that *none* of the business models will have strong fit with the company's existing skills, capabilities and/or resources. It is possible that the CEO will be choosing among business models that have only moderate or poor fit with the company's existing skills. For example, the figure that follows shows the potential fit of different business models skills, capabilities and resources of a traditional European electricity utility considering several solar power business models. The range of whites and grays (i.e., no solid blacks with strong fit) means that this is a situation in which moves by the CEO will require unflinching commitment in order to succeed. To maximize the likelihood of success, it is important for the CEO to gain insight into the market early, develop an actionable plan quickly and adjust that plan to new circumstances as they evolve.

The CEO may decide to prioritize a specific business model even without strong fit of business model with the company's skills, capabilities and resources. Why would a CEO do this? Because the opportunities (upside) from acting are large and the risks (downside) of not acting are large. This is similar to families departing Old World Europe for New World America in the early 1800s on creaky ships without a clear path to success once they reached the New World. This is similar to an individual executive leaving a secure job in a company perceived to be heading toward bankruptcy to be in a less secure job in startup. This is similar to an investor buying equity in a new company in a new industry on the unproven promise of massive potential returns. **This is all about risk and reward. Yes,**

SCREEN OF BUSINESS MODEL FIT WITH COMPANY »
(EXAMPLE: EUROPEAN ELECTRICITY UTILITY IN SOLAR POWER)

Company	Fit with sector profit pool	Risk under different scenarios	Fit with CEO interests/ priorities	Fit with company skills	Fit with other company capabilities	Fit with company resources	Overall
Polysilicon/wafer company	●/●	●/●				●/●	●/●
Chinese c-Si module company	●	●/●	●			●	
Integrated European module company	/●	/●	●/●		●/●	●/●	●
New entrant CIGS company			●/●		/●	●/●	●/●
c-Si equipment company						●/●	
European downstream company	●	●	●	●	●	●	●/●
U.S. downstream company	●	●	●	●	●	●/●	

● Strong fit   ● Limited fit   ● Little/no fit

SOURCE: PHOTON Consulting, LLC. NOTE: All data are rough estimates.

KEY TAKE AWAY » Due to massive changes within the industry, it is possible that all business models will have only moderate or poor fit for a company's existing skills.

there is big risk from the move. But there is also very large potential reward and there is significant risk from not moving fast enough from the company's existing position. In a world of "explosive growth," the opportunities to ride a powerful rising wave are exciting yet scary, but the risks of being crushed by a descending tidal wave are even scarier. This is a moment of instinctual fight or flight.

As the CEO prioritizes a specific business model, the characteristics of a flexible thinker suggest that there will be continuing reevaluation even while taking action. This reevaluation may take the form of looking at case studies of other companies that have made similar moves, more carefully assessing the competitive landscape (i.e., looking more deeply at the competition) and testing again against other screens. Given the magnitude of the change, it seems reasonable that the CEO may still look for possible reasons to exit the new business model even after starting to act toward that new business model.

Quickly, though, the action plan for the company will require clear commitment to the new course and very aggressive action. The CEO will have four main levers to consider in the pursuit of massive value creation. These four levers of value creation are:

- ○ Increasing volume
- ○ Decreasing cost

- ○ Increasing price
- ○ Decreasing risk

The CEO needs a high-level calendar for what the company is trying to achieve with each of these levers. It is useful to have a quarterly outlook at this level so that the company functions can begin planning more specific actions. For example, actions the CEO wants to take in 1Q of Year 3 may require human resource moves to begin recruiting by 1Q of Year 2 so that appropriate talent is hired by and "on-boarded" into the company. The following table is an example of an action plan. The CEO should fill in each specific action to take and the quantification of the result from these actions.

ACTION PLAN CALENDAR AND EXPECTED RESULTS »
(ILLUSTRATIVE EXAMPLE)

	YEAR 0	YEAR 1				YEAR 2				YEAR 3			
		1Q	2Q	3Q	4Q	1Q	2Q	3Q	4Q	1Q	2Q	3Q	4Q
INCREASE VOLUME	What action?	What action?	What action?	What action?	What action?	What action?	What action?	What action?	What action?	What action?	What action?	What action?	What action?
INCREASE PRICE	What action?	What action?	What action?	What action?	What action?	What action?	What action?	What action?	What action?	What action?	What action?	What action?	What action?
DECREASE COST	What action?	What action?	What action?	What action?	What action?	What action?	What action?	What action?	What action?	What action?	What action?	What action?	What action?
DECREASE RISK	What action?	What action?	What action?	What action?	What action?	What action?	What action?	What action?	What action?	What action?	What action?	What action?	What action?
OPERATING PROFIT EXPECTED ($Mn)	Estimate of impact	Estimate of impact	Estimate of impact	Estimate of impact	Estimate of impact	Estimate of impact	Estimate of impact	Estimate of impact	Estimate of impact	Estimate of impact	Estimate of impact	Estimate of impact	Estimate of impact
LEVEL OF RISK	Estimate of impact	Estimate of impact	Estimate of impact	Estimate of impact	Estimate of impact	Estimate of impact	Estimate of impact	Estimate of impact	Estimate of impact	Estimate of impact	Estimate of impact	Estimate of impact	Estimate of impact

SOURCE: PHOTON Consulting, LLC.
NOTE: All data are rough estimates.

KEY TAKE AWAY » Three-year action plan to drive value for the company.

The CEO will be trying to maneuver these four levers (increase volume, increase price, decrease cost, decrease risk) in a manner that drives the most value. In order to do this, the CEO needs an action plan that is very specific for the specific people running the specific functions of the company. Members of the team working in the functions of strategy, finance, procurement, operations, marketing, sales, HR and R&D (among others) will need very clear action plans that tie back to these four levers. An example of the type of action plan necessary is provided on pages 154–155. This action plan will need to be further refined and applied to a rollout calendar covering (most importantly) the immediate months and quarters, but also the remainder of the next 3 years.

It is important to emphasize that there is track record for value creation in distributed infrastructure that rides on the positive side of negative network effects. Generating significant value over the next 1,000 days (value at a pace commensurate with the Rockefeller-esque opportunities at hand) will require deep dedication, relentless effort and strong will, but it is plausible. For example, look at the next figure showing what numerous solar power companies have achieved over the ~1,000 days from the end of 2007 to the end 2010. This group of companies *averaged* 6X expansion from $1.1 billion in operating profit in 2007 to $6.9 billion in 2010. Impressive growth for a ~1,000 day period.

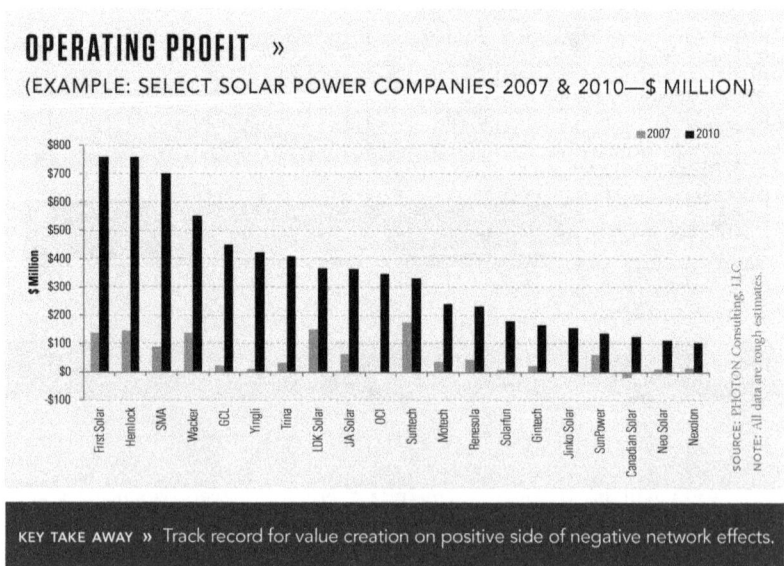

OPERATING PROFIT »
(EXAMPLE: SELECT SOLAR POWER COMPANIES 2007 & 2010—$ MILLION)

KEY TAKE AWAY » Track record for value creation on positive side of negative network effects.

SOURCE: PHOTON Consulting, LLC.
NOTE: All data are rough estimates.

152

Value creation is not easy, but it is straightforward. If a company expands volume rapidly, prevents significant price erosion and reduces costs significantly, it will generate much more profit. In the ~1,000 days from 2007 to 2010, this was clearly the case for First Solar, which saw operating profit expand more than 5X from $137 million in 2007 to $759 million in 2010.[3] A similar pattern was achieved by SMA, Wacker (Polysilicon Division) and other solar power companies.[4]

OPERATING PROFIT »
(EXAMPLE: SELECT SOLAR POWER COMPANIES 2007 TO 2010—$ MILLION)

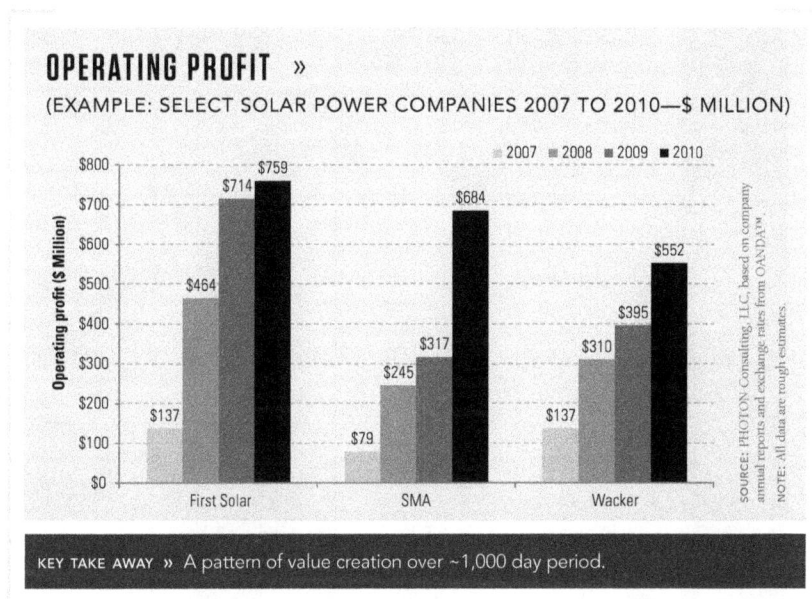

KEY TAKE AWAY » A pattern of value creation over ~1,000 day period.

Negative network effects create massive, positive opportunities for the sector I follow most closely (solar power) to expand rapidly over the next 1,000 days. Many (though certainly not all) solar power companies are well positioned with 1,000 day plans that have clear answers to the questions, "Where is the value?" and, "How can I capture it?" On the flip side, negative network effects are already causing pain for traditional electricity companies in some wealthy zip codes around the world by making the volume, price and cost of traditional electricity much less predictable. Effectively, the volume and profit growth for traditional

1,000 DAY ACTION PLAN »
(EXAMPLE: IMMEDIATE ACTIONS WITHIN FUNCTIONS OF THE COMPANY)

Core recommendations			
Volume up (Sales, MW)	Price up ($/kg, $/W, $/kWh)	Cost down ($/kg, $/W, $/kWh)	Risk down
Strategy Upstream/midstream companies: Closely partner with strong, financially stable, low-cost or differentiated customers Downstream companies: Value quantify, test and act on PV+ business models Diversify market exposure If not downstream, make downstream push through partnership or acquisition in 1H11 to protect throughput, price stability and channels directly to the end-customer	Anticipate component price declines in 2012 and aggressively pursue product/service differentiation strategies to maintain ASP premium and achieve customer lock-in	Shift toward flexible, variable cost production model by lining up contract manufacturing. Companies with proprietary/non-standard production processes should focus on low cost location capacity and improving conversion efficiency and equipment uptime Reduce fixed costs for strategy team, advisors and non-essential functions	Conduct rigorous scenario-planning based on industry and macro factors (e.g., See Solar Annual 2010 Part 1: Place Your Bets), then quantify business impact and use as the basis for strategic decision-making. Monitor interest rates, exchange rates and wholesale electricity prices in key PV demand markets
Finance Offer/arrange system financing from financial partners, investors and/or balance sheet Downstream companies: Develop expertise and track record to pool projects for financing/ capital raising	Bankability will be an essential criteria for pricing strength in 2012-2013, so prioritize bankability in anticipation of 2012: Balance sheet strength/cash preservation, component insurance, cross-branding with stronger brands, track record building, independent product performance studies	Leverage innovative structures (examples: Municiple bonds, term debt, SREC off-take agreements), strong relationships and track record to drive down project development cost of capital	Aggressively build up cash position, secure credit and reduce balance sheet leverage in anticipation of challenging 2012-2013 environment Arrange lines-of-credit and other flexible, readily-accessible sources of working capital in 1H11
Capex Identify attractive candidates and/or distressed assets for capacity acquisition and new market presence/exposure after crash commences when other companies need to raise cash Meet near-term volume needs via contract manufacturing wherever possible rather than internal capex	Focus capex in areas that enable higher price such as capex for higher degree of downstream vertical integration (to capture higher $/W) or segments within manufacturing that have may be able to retain price premium (e.g., super-mono)	Avoid bringing incremental internal capacity/production online unless it can be fully ramped/operational by end of 1Q11	Limit near-term capex spend to preserve cash and financial flexibility during 2012-2013 sector downturn—valuable distressed asset options may emerge Time new internal capex for 2Q13-3Q13
Procurement Plan procurement for sector downturn (i.e., less volume) but include terms in procurement contracts for feedstocks that have options for higher volume deliveries in case customer demand is higher than expected	Plan procurement during downturn of high value-added feedstocks that enable higher price products	Avoid prepayments on long-term feedstock/purchase contracts, even at the expense of higher average purchase price in out years. Avoid take-or-pay contracts and reserve the option for spot purchases in 2012. Attempt to peg feedstock contracts to industry indices but with caps on upside in the procurement price	Avoid prepayments on long-term feedstock/purchase contracts and reset pricing on feedstock procurement contracts quarterly or monthly. Value long-term relationships small number of key, strong, trusted customers

KEY TAKE AWAY » Final step is a clear action plan for each function of the company.

Core recommendations			
	Volume up (Sales, MW)	Price up ($/kg, $/W, $/kWh)	Cost down ($/kg, $/W, $/kWh)	Risk down
Operations	Plan operations for sector downturn (i.e., less volume) but include flexible back-up operations plans for higher volume production in case customer demand is higher than expected	Contract manufacture modules in attractive emerging markets to accelerate time to market and address local content requirements where applicable	Upstream: Controlled expansion, cost reduction, technology leadership, operational excellence Midstream: Work with—not against—low cost Asian manufacturing; attempt to shift toward higher variable cost production model Downstream: Assemble BOS pre-fab/off-site, use technology/logistic/operational approaches to accelerate project buildcycle/lifecycle	Focus on supply chain management optimization to minimize inventory days and reduce working capital usage
Marketing	Identify, prioritize, focus on brand and channels across a diverse group of emerging markets with lower saturation risk (examples: Australia, India, UK, Ontario, New Jersey (US), Arizona (US), Texas (US), Massachusetts (US), Pennsylvania (US)	Identify and target higher (electron) value markets and market segments (particularly residential/small commercial)	Consider reduction in marketing expenses (e.g., advertising, brand building) during period of downturn; tie marketing costs more closely to volume and/or price performance	Apply systematic, risk-based screening criteria customer/partner selection. Criteria should include: (1) Market saturation risk exposure, (2) market disruption hedging capacity, (3) financial/balance sheet health, (4) cost leadership, (5) defensible channels to market
Sales	Reduce sales % allocation to Germany and other markets with elevated near-term demand slow-down risk (France, Belgium, Slovakia). Focus on opening new/emerging markets outside of Europe Develop differentiated offerings/value propositions Develop strong pre- and after-sales support (system design, financing, lead generation, training, warranty support) to strengthen and differentiate relationships with customers	Identify and target higher (electron) value markets and market segments (particularly residential/small commercial)	Consider reduction in sales expenses (e.g., fixed legal and sales expenses) during period of downturn; tie sales costs more closely to volume and/or price performance Carefully analyze the risks (policy, transmission and distribution infrastructure, exchange rate, etc.) of entering a new market prior to investing in/establishing sales presence	Pursue more prepayments on long-term sales contracts, even at the expense of ASP in out years Tighten A/R policies and procedures Pricing premia expected to narrow in over-supply situation—identify/focus on low-cost customers with sustainable and defensible channels to market Value long-term relationships small number of key, strong, trusted customers
R & D	Evaluate ways to boost throughput from existing capacity	Evaluate ways to improve products and services in areas that provide more defensible pricing (e.g., conversion efficiency increases, PV+ kits, AC modules, after-sale services, O&M/performance prediction)	Cell/module: Evaluate options for using lower cost inputs and ways to boost throughput from existing capacity BOS: Pre-assembled components, "BOS in a box" solutions, string design/electrical system optimization	Maintain controlled R&D budget through the crash with focus on key market opportunities (examples: PV+, BOS optimization, increased cell/module integration through back-side contacts, wafer thickness reduction)

electricity has already shifted from slightly positive growth to negative growth in some wealthy zip codes, and it appears likely to quickly do the same in a much broader range of wealthy regional geographies. CEOs facing this downside need to quickly acknowledge their situation, procure the right ingredients discussed in the previous chapter and develop actionable answers to the questions, "Where is value?" and "How to capture it?"

But the pattern of benefiting from the positive sides of negative network effects is not confined to solar power. Other sectors are poised for rapid expansion of their profit pools in the coming 3 years. This includes sectors close to solar power, such as storage, other distributed electricity generation, distributed electricity controls and energy efficiency. In addition, there is a strong case to be made that negative network effects will create big opportunities in the food, water, communications, health care and other sectors. Specific industries that face game-changing shifts with massive potential up- and downside include:

○ Agriculture	○ Automotive
○ Banking	○ Brokerage
○ Chemicals	○ Consulting
○ Consumer products	○ Department stores
○ Electronics	○ Energy
○ Executive search	○ Financial services
○ Food and beverage	○ Grocery
○ Health care	○ Investment banking
○ Manufacturing	○ Pension funds
○ Private equity	○ Publishing
○ Real estate	○ Restaurants
○ Retail & wholesale	○ Securities & commodity exchanges
○ Services	○ Technology
○ Telecommunications	○ Transportation
○ Water	○ Venture capital

DISTRIBUTED INFRASTRUCTURE COMPANY ILLUSTRATIVE
INCOME STATEMENT » (ILLUSTRATIVE EXAMPLE—2011–2014)

	Year 0 2011	Year 1 2012	Year 2 2013	Year 3 2014
Volume (million units)	200	500	1100	1800
Average price ($/unit)	$2.50	$2.40	$1.90	$1.30
Revenue ($ Million)	$500	$1,200	$2,090	$2,340
Cost ($/unit)	$1.80	$1.50	$1.30	$0.90
Cost ($ Million)	$360	$750	$1,430	$1,620
Operating profit ($ Million)	$140	$450	$660	$720

SOURCE: PHOTON Consulting, LLC.
NOTE: All data are rough estimates.

KEY TAKE AWAY » Opportunities to rapidly build small and mid-size existing companies within 1,000 days.

In all of these (solar power sector, sectors close to solar power and other sectors as diverse as food, water, communications and health care), I am convinced that there are opportunities to rapidly build small and mid-sized existing companies to reach hundreds of millions of dollars in operating profit within 1,000 days with the potential to expand quickly beyond that time. My hope is that CEOs will follow Rockefeller's advice about not being afraid to "give up the good to go for the great."

⌃

BIRTH AND DEATH
OF INDUSTRIES

The release of atomic energy has not created a new problem. It has merely made more urgent the necessity of solving an existing one.
—*Albert Einstein (1879–1955)*

BIRTH AND DEATH

This book began with the hypothetical question, "What would you do if a pregnant woman went into labor in the back of your taxi during rush hour?" My guess is that it would be an incredible honor to participate if all ends well, but a true horror if anything goes wrong. In my professional life, I am watching the birth of new industries and living with a deep tension between the magical beauty of potential birth and the tragic worry of potential death.

This tension is a daily part of my professional life. Birthing new industries is thrilling. It is with awe that I watch the spectacle of companies growing exponentially, costs dropping quickly and customers adopting rapidly. The pace and scale I am observing within my daily business activities reminds me of a line written by poet Maya Angelou and read at the 50th anniversary of the United Nations: "The haughty neck is happy to bow. And the proud back is glad to bend." I am genuinely humbled to be an observer of this inspirational transformation.

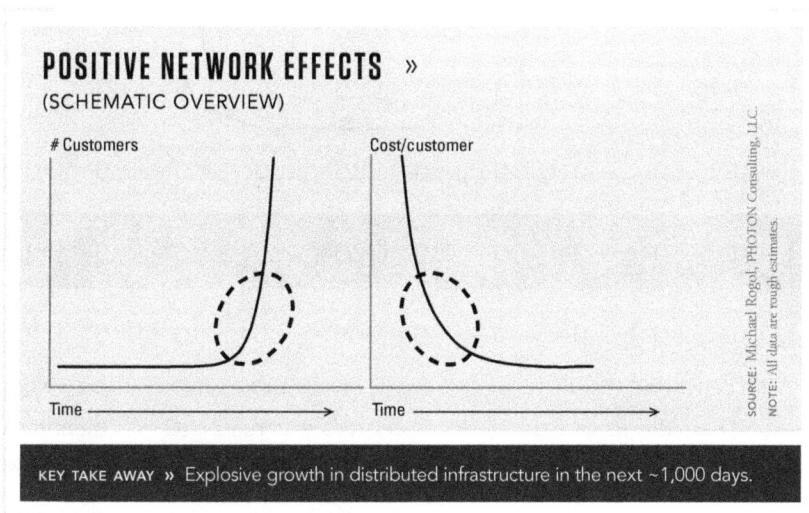

POSITIVE NETWORK EFFECTS »
(SCHEMATIC OVERVIEW)

Customers

Time

Cost/customer

Time

SOURCE: Michael Rogol, PHOTON Consulting, LLC.
NOTE: All data are rough estimates.

KEY TAKE AWAY » Explosive growth in distributed infrastructure in the next ~1,000 days.

Yet the emergence of distributed infrastructure creates an urgent need to find solutions for traditional infrastructure. While the wealthy residents in the wealthy zip codes of the world will surely find ways to maintain their quality of life, the same will not be true for their less-wealthy neighbors. As the wealthy have adopted distributed infrastructure, the traditional infrastructure networks are *already* unable to maintain volume growth and costs are *already* rising rapidly. The rapid growth trajectory of distributed infrastructure means a larger headache, then migraine and then aneurysm for traditional infrastructure networks in wealthy zip codes. For those with the resources to afford distributed infrastructure, the increasing level of self-sufficiency and generation-near-the-point-of-use creates a level of safety and security. But those without the resources to afford distributed infrastructure are left relying on centralized infrastructure networks that quickly become less reliable and more expensive.

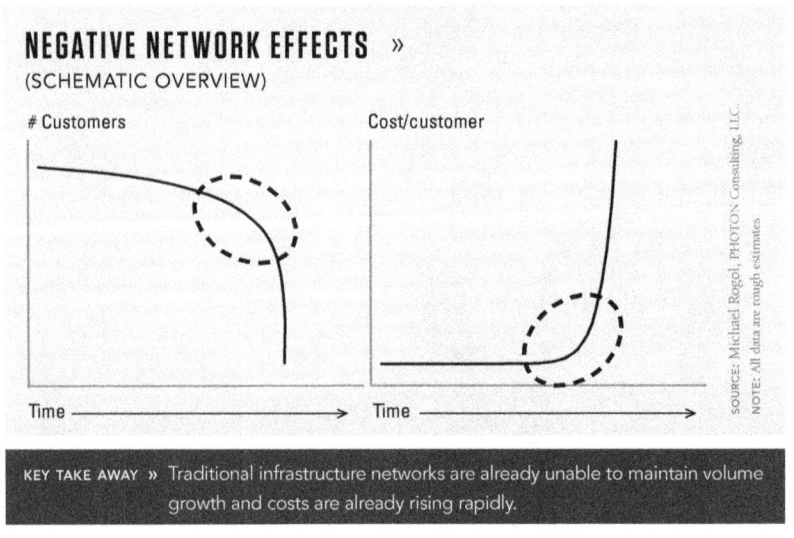

NEGATIVE NETWORK EFFECTS »
(SCHEMATIC OVERVIEW)

Customers

Cost/customer

Time

Time

SOURCE: Michael Rogol, PHOTON Consulting, LLC.
NOTE: All data are rough estimates

KEY TAKE AWAY » Traditional infrastructure networks are already unable to maintain volume growth and costs are already rising rapidly.

I am convinced that there are massive downside risks from negative network effects. These downside risks are for companies that run or rely on traditional infrastructure networks, including our electricity, transportation, communication, food, water and health care networks. These risks are perhaps even more pressing for individuals and families, especially those at the lower end of the socioeconomic spectrum. Together, this situation will drive personal-level, business-level, industry-level and societal-level challenges. To paraphrase Einstein, the emergence of negative network effects from the growth of distributed infrastructure is not creating new problems. It is merely making the necessity of solving existing ones more urgent. To quote Lincoln, we cannot escape the responsibility of tomorrow by evading it today.

In conclusion, the upsides and the downsides from negative network effects are poised to transform industries over the next 3 years. The explosive growth of distributed infrastructure will reshape energy and infrastructure with a boom in the next 1,000 days, echoing through our economy for decades to come.

ACKNOWLEDGMENTS

⌃

My wife Susan brought the idea of this book to life. She inspired me to move my attention beyond energy, encouraged me to dig deeper on topics about which she knew much and I knew little and pushed me to prioritize this book over sleep and many other activities. Without her selfless support, impactful insights and razor sharp edits, *Explosive Growth* would never have ignited. I may never be able to thank her enough for all her contributions, but I look forward to reciprocating. Similarly, we together are forever grateful to our parents and grandparents for pushing us to live life to the fullest while giving to others.

Frank Britt's curiosity also drove this book. He has relentless curiosity and poses the question, "Why?" with such passion that the room seems to shake from the force of a marching band during half time at the Rose Bowl as his eyes shift endlessly looking for another intellectual toy to occupy a brain that races at NAS-CAR pace already anxious to scream out of the next pit stop wearing fresh tires. I am grateful for his friendship, his partnership and his ongoing commitment to explosive growth. I am also grateful for this process, which enabled me to visit regularly with his incredible family.

Many other people supported and guided me through this writing effort. I am grateful to all peer reviewers and editors, including John Damon, Maryanne Grady, Mark Farber, Ramsay Stirling and David Supan.

Joe Lara's tireless leadership in business and unflappable camaraderie in life make him the ultimate professional partner. I thank him for rowing on both sides of our business' boat while I jumped into the water, for his coaching and for his unshakable commitment to finding the right answer. Without him, this book might still exist, but the remainder of my professional life would be in shambles. I am excited to get back to rowing with you and the rest of the PC team.

Many people supported and guided me through this research effort. Specifically, I am grateful to all members of PHOTON Consulting including founders Mark Farber, Joe Lara, Martin Meyers, Chris Porter, Josh Rogol, and Joonki Song plus Rob Blumberg, Chris Bolman, Jason Cooper, Stephanie Curtis, David Southwick and Ramsay Stirling who perpetually shape my thinking about solar power, energy, economics, business and professional friendship. I appreciate the energetic dedication of Ted Feierstein, Alex Fraenkel, Nana Hori, Nicholas Lara, Ravi Manghani, Ben Merewitz, Christina Peaslea, Michael Shields, Rob Trangucci and all others in the next generation of rising stars within the company. I have deep gratitude to Lucy Baez, Kara Berrini, Julie Morneau, Marcia Regan, Scott Rubinstein and Phil Tracy for their incredible efficiency in keeping me on track and their dedicated professionalism. I also applaud the team at Greenleaf Book Group that is remaking the world of publishing and Newman Communications (particularly David Ratner) for all the help getting this project into the capable hands at Greenleaf.

I thank the tens of thousands of clients that support PHOTON and the thousands of companies that are building deep relationships with PHOTON Consulting. In particular, I am grateful to "A" and "S" for being a flexible thought partners and mentors who push my thinking and inspire the best from all around them. I also thank "SK" for many years of friendship, mentorship and introductions to incredible people.

I thank Professor Ernest Moniz, Professor Rebecca Henderson, Professor Emanuel Sachs, Professor Richard Lester and Professor David Marks for their countless hours of mentoring as members of my PhD committee and for their understanding of the personal reasons that prevented me from completing the dissertation; Travis Franck, Kate Martin and David Miller for their feedback and suggestions to improve my early research at MIT; the Martin Fellows Program, Shell Oil, GE Energy Financial Services and CLSA for their funding and financial support of this research; Thomas Seitz, Dominic Barton, Joseph Stanislaw, Daniel Yergin, Kevin Lindemer, William F. Martin and the many other energy mentors who have spent two decades imparting energy sector wisdom.

I also thank my entire family—Abby, Arlene, Grandma Boone, Grandma Rogol, Hannah, Jennifer, Johnnie, Josh, Karen, Kate, Laura, Mark, Maritucker, Neal, Olivia, Paul, Rob, Sarah, Tucker, Wong and again our parents—for providing endless encouragement.

Finally, I pray that the next generation, including Meyer Rays, will have strong infrastructure to support a bright future.

NOTES

⌃

INTRODUCTION

1 For additional information see the proceedings of PHOTON's Solar
   Electric Utility Conferences. As examples, see Walter Albrecht and
   Peter Schwaegerl, "Solar Impact on Distribution Grids," http://www.
   photon-expo.com/en/sthcs_2011_europe/conferences/seuc_program.
   htm and Rainer Bäsmann, "Challenges of Handling Massive PV Installations—View from a Grid Operator," http://www.photon-expo.com/en/
   pts_2010_europe/seuc_2010_program.htm.

2 Historical data prior to 2000 from "Annual Survey of Solar Power Manufacturing" in the annual newsletter from Paul Maycock, *PV News* 1999
   and 2000; Data from 2000 from PHOTON Consulting as published
   in yearly Solar Annual reports. For example, see "Solar Annual 2010-
   2011: Cash-In" (PHOTON Consulting, 2010); "Solar Annual 2009: Total
   Eclipse" (PHOTON Consulting, 2009); "Solar Annual 2008: Four Peaks"
   (PHOTON Consulting, 2008); "Solar Annual 2007: Big Things in a Small

Package" (PHOTON Consulting, 2007), "Solar Annual 2006: The Gun Has Gone Off" (PHOTON Consulting, 2006).

3 European Energy Exchange, www.eex.com, July 22, 2011 compared to July 23, 2010. Data for natural gas, coal and electricity prices for Cal-13.

CHAPTER 1

1 United States Postal Service, *2010 USPS Annual Report* (Washington, DC: USPS, 2010), www.usps.com/financials/_pdf/annual_report_2010. pdf; Kyle Mudry and Justin Bryan, "Individual Income Tax Rates and Shares, 2005," *Statistics of Income Bulletin*, Winter 2008, www.irs.gov/ pub/irs-soi/05inrate.pdf. Data from recent years indicates annual taxable returns of 83 to 97 million per year, with an average of roughly 90 million per year. See page 2 for details on Michael Parisi, "Individual Income Tax Returns, Preliminary Data, 2008," *Statistics of Income Bulletin*, Winter 2008, http://www.irs.gov/pub/irs-soi/10winbul.pdf and Adrian Dungan and Kyle Mudry, "Individual Income Tax Rates and Shares, 2007," *Statistics of Income Bulletin*, Winter 2008, http://www.irs.gov/pub/irs-soi/10winbul.pdf.

2 There are many analyst reports and interviews of USPS executives available online. One example is Boston Consulting Group, "Projecting US Mail Volumes to 2020: Final Report—Detail," March 2, 2010, *about.usps. com/future-postal-service/bcg-detailedpresentation.pdf.*

3 *The United States Postal Service: An American History 1775–2006* (Washington, DC: USPS, 2006) www.usps.com/history/anrpt01/.

4 "Online Bill Payments Surpass Checks for the First Time Among Internet-Connected Households," Check Free Corporation, April 12, 2007, http://www2.prnewswire.com/cgi-bin/stories.pl?ACCT=104&STORY=/www/story/04-12-2007/0004564261&EDATE.

5 United States Postal Service, *2010 USPS Annual Report* (Washington,

DC: USPS, 2010), www.usps.com/financials/_pdf/annual_report_2010.pdf and similar statements by Postmaster General/CEO Patrick Donahoe before the Committee on Homeland Security and Governmental Affairs of the U.S. Senate on September 6, 2011, http://about.usps.com/news/speeches/2011/pr11_pmg0906.pdf.

6 There are a variety of sources that estimate the size, number, value, and patterns of bill payments, including online bill payments. For example, see "Survey Proves Value of Online Banking, Bill Pay," *The Financial Brand.com*, June 4, 2010, http://thefinancialbrand.com/12000/fiserv-2010-online-banking-bill-pay-trends-report; Kirk Gripenstraw, "Online Bill Pay Longevity and Lifetime Value Study" (Chicago, IL: Aspen Marketing Services, 2009) www.fiserv.com/WP_aspen-bill-pay-value2009.pdf; Sharon Brant, "Consumer Payment Preferences for Bill Payment," (Atlanta, GA: First Data, 2008), http://www.firstdata.com/enews/CPPBrief_BillPayment.pdf; Daniel Hough, Mark Riddle, Chris Allen, and Melissa Fox, "World of Choice: Consumer Payment Preferences," *BAI Banking Strategies*, Jan/Feb 2009, *www.hitachiconsulting.com/.../AR_ConsumerPayments_AllenFox_JanFeb09_Final.pdf*; and Emmett Higdon, with Benjamin Ensor, Peter Wannemacher and Courtney Tincher, "US Online Bill Payment Forecast: 2009 to 2014" (Cambridge, MA: Forrester, 2009), http://www.forrester.com/rb/Research/us_online_bill_payment_forecast_2009_to/q/id/46859/t/2. While estimates range across a these sources, the trends are clear: Rising number of online households, rising percentages of online bill pay for online households, rising number of bills paid online per average online bill pay household, and so on.

7 This is a secondary influence mentioned in phone interviews with online bill pay analysts.

8 $67 billion revenue divided by 171 billion units of mail.

9 $0.39/unit X 28 letters/month = decrease of $129/year per upper middle-class household using online bill pay.

10 Gripenstraw, Kirk. "Online Bill Pay Longevity and Lifetime Value Study"

(Chicago, IL: Aspen Marketing Services, 2009), www.fiserv.com/WP_aspen-bill-pay-value2009.pdf.

11 This estimate comes from a variety of sources and is likely conservative. One example is Sharon Brant, "Consumer Payment Preferences for Bill Payment," (Atlanta, GA: First Data, 2008), http://www.firstdata.com/enews/CPPBrief_BillPayment.pdf, which estimates 30 million direct-billed online bill payment households and an additional 23 million third-party online bill payment households.

12 30 million online bill customers is a big number. Researchers have different accounting for the number of online bill pay customers. For example, estimates for 2008 range from roughly 30 million customers to roughly 50 million customers depending on the source of information and what is being counted. As an example of one study, see Daniel Hough, Mark Riddle, Chris Allen, and Melissa Fox, "World of Choice: Consumer Payment Preferences," *BAI Banking Strategies*, Jan/Feb 2009, www.hitachiconsulting.com/.../AR_ConsumerPayments_AllenFox_JanFeb09_Final.pdf.

13 One peer reviewer asked, "Why was 2008 a loss but 1996 profitable?" Basically, why did USPS lose money with 203 billion pieces in 2008 but made money with 183 billion pieces in 1996? The simple answer is that fixed costs continued to rise in the USPS network from 1996 to 2008 because delivery points rose from 128 million in 1996 to 151 million in 2008. United States Postal Service, "Delivery Points 1905 to 2010, in millions," (Washington, DC: USPS, January 2011), http://about.usps.com/who-we-are/postal-history/delivery-points-1905-to-2010.pdf.

14 My best guess of a plausible range of USPS mail volume lost to online bill pay is between 7 billion and 15 billion pieces based on a review of a variety of analyses of online bill pay trends and interviews with experts from the USPS and consulting firms serving USPS.

15 Emmett Higdon, with Benjamin Ensor, Peter Wannemacher and Courtney Tincher, "US Online Bill Payment Forecast: 2009 to 2014"

(Cambridge, MA: Forrester, 2009), http://www.forrester.com/rb/
Research/us_online_bill_payment_forecast_2009_to/q/id/46859/t/2.

16   General Accounting Office's *High Risk List* including *Restructuring the U.S. Postal Service to Achieve Sustainable Financial Viability.* See http://www.gao.gov/highrisk/risks/efficiency-effectiveness/restructuring_postal.php.

CHAPTER 2

1   Historical data prior to 2000 based on "Annual Survey of Solar Power Manufacturing" in the annual newsletter from Paul Maycock, *PV News*, 1999 and 2000.

2   Ibid; Data from 2000 based on PHOTON Consulting and available in yearly Solar Annual report series. For example, see: "Solar Annual 2010–2011: Cash-In" (PHOTON Consulting, 2010); "Solar Annual 2009: Total Eclipse" (PHOTON Consulting, 2009); "Solar Annual 2008: Four Peaks" (PHOTON Consulting, 2008); "Solar Annual 2007: Big Things in a Small Package" (PHOTON Consulting, 2007), "Solar Annual 2006: The Gun Has Gone Off" (PHOTON Consulting, 2006).

3   I have been looking for other examples of sectors that have grown at a 45% compound annual rate over a 15-year period, but have not yet found any. The sectors in which I have looked include railroads, oil, natural gas, information technology, computer hardware, software, telecommunications and pharmaceuticals. There *are* examples of companies growing this fast for this long, but I have not been able to identify any sectors that have grown at this high rate for a 15-year period. I appreciate all suggestions in this area to identify potential case studies that are applicable.

4   "Solar Annual 2009: Total Eclipse," http://www.photonconsulting.com/solar_annual_total_eclipse.php.

5   "Solar Annual 2008:Four Peaks," http://www.photonconsulting.com/solar_annual_2008.php.

6   PHOTON Consulting's *The Wall* data platform.

7   Ibid; International Energy Agency (IEA) *Electricity Information 2010* for electricity sector data, http://www.iea.org/publications/free_new_Desc. asp?PUBS_ID=2036.

8   In 2009, we led a group of U.S. utility executives to Spain as part of a fact-finding mission organized by the U.S. Solar Electric Power Association (SEPA). This fact-finding mission was designed to help share the challenges of solar power's rapid growth observed in Spain with a couple dozen U.S. utilities. The results from the fact-finding mission (including the size of the gap in Spain's electricity sector) were then synthesized and shared with hundreds of U.S. utilities.

9   Ibid.

10   The World Bank, "World Development Indicators 2011," http://www. google.com/publicdata?ds=wb-wdi&met=sp_pop_totl&idim=country:E SP&dl=en&hl=en&q=population+of+spain.

11   PHOTON Consulting's *The Wall* data platform.

12   Ibid.

13   Ibid.

14   Lars Paulsson, "Solar Doubling Drives Down German Power Prices: Energy Markets," *Bloomberg*, Sept. 22, 2010, http://www.bloomberg.com/ news/2010-09-21/solar-doubling-gas-glut-drive-down-german-power-prices-energy-markets.html. This is an example of solar power driving down wholesale power prices. This occurs because there is more supply of electricity with the addition of solar power. In contrast to the wholesale power price, the retail power price is going up in Germany as the result of policies that spread the cost of solar power incentives across all end-customers. This contrast of decreasing wholesale power prices and rising retail power prices is one anomaly that makes the impact of solar power on electricity difficult to understand for many observers who are not expert in analyzing electricity markets.

15 Rough estimate based on data from International Energy Agency (IEA) *Electricity Information 2010.*

16 PHOTON Consulting's *The Wall* data platform.

17 The World Bank, "World Development Indicators 2011," http://www. google.com/publicdata/explore?ds=d5bncppjof8f9_&met_y=sp_pop_tot l&idim=country:DEU&dl=en&hl=en&q=population+data+germany.

18 PHOTON Consulting's *Solar Monthly* from 2011 and *Electricity Discussion* from July 2011, http://www.photonconsulting.com/consulting_services.php.

19 Getting better information and having genuine debate is the central reason we started PHOTON's Solar Electricity Conference series. For an example of previous conference agendas, see Photon's 6th Solar Electric Utility Conference program, Berlin, April 14, 2011, http://www.photon-expo. com/en/sthcs_2011_europe/conferences/seuc_program.htm; Photon's 2nd Solar Electric Utility Conference program, Munich, March 6, 2009, http://www.photon-expo.com/en/pts_2009_europe/pvseuc_2009_pro- gram.htm; and Photon's 3rd Solar Electric Utility Conference Program, San Francisco, February 4, 2010, http://www.photon-expo.com/en/ pts_2010_usa/seuc_2010_program.htm.

20 PHOTON Consulting's *The Wall* data platform for solar power data.

21 Population data from The World Bank, "World Development Indicators, 2011," http://www.google.com/publicdata/explore?ds=d5bncppjof8f9_&& met_y=sp_pop_totl&idim=country:CZE&dl=en&hl=en&q=populatio n+of+czech+republic#ctype=l&strail=false&nselm=h&met_y=sp_pop_ totl&scale_y=lin&ind_y=false&rdim=country&idim=country:CZE&ifdi m=country&hl=en&dl=en.

22 PHOTON Consulting's *The Wall* data platform for solar power data. Inter- national Energy Agency (IEA) *Electricity Information 2010* for electricity sector data.

23 Ladka Bauerova "Czech Solar Power Risks Industrial Power Price Surge," *Bloomberg*, Sept 17, 2010, http://www.businessweek.com/news/2010-09- 17/czech-solar-power-risks-industrial-power-price-surge.html. This is the

counterside to the German example. This is an example of solar power driving up retail electricity prices. Per previous note, this occurs due to government policies that spread the cost of solar power subsidies across end-customers. In contrast to retail prices, wholesale electricity prices are pushed down by the additional supply from solar power.

24 PHOTON Consulting's *The Wall* data platform.

25 Ibid.

26 Ibid.

27 Ibid.

28 Ibid; In 2010, the German traditional utilities were made "whole" by Government policy, but focusing on this is like saying, "The bandage makes the wound irrelevant."

29 Much research has been conducted on the lifetime of solar power systems. Nearly all estimates for crystalline silicon solar power systems are more than 20 years old, with many above 25 years and some above 30 years. For one example, see P. Frankl, A. Masini, M. Gamberale and D. Toccaceli, "Simplified Life-cycle Analysis of PV Systems in Buildings: Present Situation and Future Trends", *Progress in Photovoltaics: Research and Applications*, 6(2), pp. 137–146, 1998.

30 Data on solar power installations PHOTON Consulting's *The Wall* data platform. Additional information available from Bundesnetzagentur (government reporting agency responsible for collecting and disseminating data on solar power), http://www.bundesnetzagentur.de/enid/525ccad2 70251bc90c90a22ace3de32e,0/Erneuerbare_Energien_Gesetz__EEG_/ Verguetungssaetze_Photovoltaikanlagen_5y2.html; Data on electricity and electricity capacity additions from International Energy Agency (IEA), including *Electricity Information 2010* for electricity sector data, http://www.iea.org/publications/free_new_Desc.asp?PUBS_ID=2036.

31 Based on data from the European Energy Exchange (www.eex.com) and confirmed by numerous interviews with utility executives, regulators and policy makers in Germany.

32 Ibid; In 2010, the German traditional utilities were made "whole" by Government policy, but focusing on this is like saying, "The bandage makes the wound irrelevant."

33 PHOTON Consulting's *Solar Monthly* from 2011 and *Electricity Discussion* from July 2011, http://www.photonconsulting.com/consulting_services.php.

34 In the U.S. Postal Service example, a half-empty USPS truck still operates well. In contrast, many electricity assets have operating challenges (in addition to economic and financial challenges) at low utilization rates.

35 All data are rough estimates. The graphic presents the cumulative value of financed solar electric systems since the inception of SolarLease by SolarCity in 2008. Data for 2011 and 2012 are estimates based on publicly available data from SolarCity. For example, as of June 2011, Solar-City had more than $950 million of deals in progress, with expectation of achieving more than $785 million of deals completed in 2011.

CHAPTER 3

1 Data from CTIA—The Wireless Association, "US Wireless Quick Facts," http://www.ctia.org/advocacy/research/index.cfm/aid/10323.

2 Positive network effects are a familiar story. There are a broad set of products that rely on networks to create value for both suppliers and users, ranging from transportation (e.g., networks of railroads, interstate highways, and airports) to food (networks of supply and distribution for chickens, sodas, hamburgers) to water (networks to deliver both tap water and bottled water) to communications (networks to deliver phone calls, email, and standard mail). Positive network effects are behind the wealth captured by business leaders as diverse as Herb Kelleher (builder of Southwest Airline's network of second-tier air hubs), Frank Purdue (builder of Purdue's network for procuring, handling, and delivering

chickens) and Fred Smith (builder of FedEx's network to deliver packages). For end-users, there are easy-to-identify benefits from these positive network effects, such as buying low-priced airline tickets (often well below the price of 40 years ago), eating a Costa Rican banana during the middle of winter in Boston (something commonplace today but a rarity 40 years ago) or receiving updates from an expanding circle of friends and family on Facebook (something not yet imagined 40 years ago). Positive network effects are so common and important in modern life that we have intimate familiarity with them, even if we do not think about them using the term "positive network effect."

3 Also known as the Pequot Trail. For an example of the technology transitions along the same basic transportation network, see Thomas Williams Bicknell, *The History of the State of Rhode Island and Providence Plantations* (New York: The American Historical Society), vol. 2. In Chapter 35 on "Roads, Post Roads, and Post Offices," Bicknell writes, "A high road or highway is a path or road slightly elevated above the land on its sides. In early Rhode Island, before the advent of horses, oxen, carts and carriages, the people traveled along the trails made by the [native Americans]. These trails, now converted in many instances into public roads, are the oldest and most permanent of the memorials left by the [native Americans] . . . For instance the Pequot Trail, taking its name from a small but active tribe in Southern Connecticut, is easily traceable . . . This trail was a trunk line between great tribes, from which diverged secondary trails . . . These primary and secondary trails covered New England with a network of well-worn lines of travel—the guides of many of our modern roads."

4 For a rich history of the early transportation network in the U.S., see Stewart Hall Holbrook, *The Old Post Road: The Story of the Boston Post Road* (New York: McGraw-Hill, 1962).

5 There is a general *perception* that technology transitions by end-users drive more volume within the existing infrastructure and that the existing infrastructure will adjust to accommodate. Today, we are observing

changes in end-user technology (e.g., rising penetration of hybrid and electric vehicle technology taking market share from combustion engine technology) at the same time that volume (e.g., passengers, vehicles, vehicle miles) increases and at the same time that the transportation network continues to be widened, lengthened and strengthened. This pattern (e.g., end-user technology transition combined with growth in volume on the traditional network combined with adaptation of the traditional network) blends together so seamlessly that we all expect this to work. When we purchase a "next generation" computer or mobile phone, we just assume it to work on the existing networks. The pattern has occurred so often and for so long that we have stopped *thinking* carefully about the underlying fundaments. It is important to revisit these fundamentals like a detective to make sure that there is not a gap between perception and reality. To avoid potential confusion, other factors also positively impacted the volume going through these networks. For example, the increasing population (demographic growth) and expanding size of the economy (economic growth) also positively impacted network volumes. The points being made in this chapter do not discuss these other factors. A more complex point than made in this chapter is that technology transitions often occur within a context of demographic and economic growth. A review of technology transitions in Italy (particularly in specific regions during specific periods of negative demographic growth and negative economic growth) may be worth pursuing in the future based on initial research there during the mid-2000s. In addition to demographic and economic factors, another added layer of complexity comes from the fact that technology transitions often require building additional infrastructure that inter-operates with existing infrastructure (e.g., mobile telecom towers connecting to the traditional telecom networks).

6   This is not one telecom provider stealing market share from another telecom provider with the end-customer still relying on telecommunications infrastructure network. This is not mobile telecom gaining customer

adoption in a pattern that captures more end-customer revenue for the mobile telecom provider instead of the fixed-line telecom provider, but the end-customer's volume still often relies on the landline telecommunications infrastructure network. This *is* online bill pay via the Internet's infrastructure network instead of physically mailing bill pay via the traditional mail delivery infrastructure network.

7   It is also *plausible* that it will not occur. The *plausibility* of negative network effects having and not having significant impact is a central reason that I am pursuing people who are proven flexible thinkers, such as Frank Britt. The characteristic of flexible thinking is important to quickly getting to the best possible answer based on available information but then also adjusting this answer over time as more information becomes available.

8   PHOTON Consulting's *Solar Monthly* from 2011 and *Electricity Discussion* from July 2011, http://www.photonconsulting.com/consulting_services.php; Data on electricity market in Germany from European Energy Exchange (www.EEX.com) plus interviews with electricity companies, traders, regulators, and policy makers.

9   All data are rough estimates. Assumes summer peak of 71GW, winter peak of 79GW, weekend peak of 52GW and annual peak of 84GW for 2010, with 2% annual growth.

10   Anne Marie Chaker, "The Science Project You Can Eat" *The Wall Street Journal*, January 26, 2011, http://online.wsj.com/article/SB10001424052 748704698004576103981377557622.html.

11   In Japan, there was a 32-fold increase (+3086%) in annual installations in the decade from 1994 to 2003, equating to 41% compound annual growth. For a detailed discussion, see my master thesis, Michael Rogol, *Why Did the Solar Power Sector Develop Quickly in Japan?* (MIT Engineering Systems Division, 2007).

12   Celia Soudry, "Kogi's Roy Choi is First Food Truck Chef Named "Best New Chef" by Food & Wine," *LAWeekly Squid Ink* blog, April 7, 2010,

http://blogs.laweekly.com/squidink/2010/04/food_wine_kogi_roy_choi_
best_n.php.

13   Data from PHOTON Consulting's *The Wall* data platform. Comment
based on many interviews in Germany and hosting of several PHOTON
Solar Electric Utility Conferences with leading German utilities attending
and presenting.

14   Anne Marie Chaker, "The Science Project You Can Eat," *The Wall Street
Journal*, January 26, 2011, http://online.wsj.com/article/SB10001424052
748704698004576103981377557622.html.

15   Magazine advertising in distributed infrastructure is something with
which I have some familiarity. My company, PHOTON, publishes a dozen
magazines in the solar power and renewable energy sector, with most of
the economics of these magazines driven by advertisements.

16   Email from executive used with permission without using name.

17   Jennifer Levitz, "Postal Service Eyes Closing Thousands of Post Offices,"
Wall Street Journal, January 24, 2011, http://online.wsj.com/article/SB10
001424052748704881304576094000352599050.html.

18   It possible that the negative impacts of negative network effects may be
avoided due to countervailing forces. In the case of mobile telecom, it
was more phone minutes per person, more phones per person, and more
phones per business that differentiated a technology transition trend
from a full-blown negative network effect. In the case of electricity, it is
possible that rising home electronics (e.g., big TVs) and rapid adoption of
electric vehicles might fuel growth (not just slower volume declines but
actual volume increases) within traditional electricity networks. This is
also plausible. If this or other similar countervailing forces were to occur,
a full-blown negative network effect might be avoided by the traditional
electricity companies and traditional electricity infrastructure.

CHAPTER 4

1   PHOTON Consulting's *The Wall* data platform.

2   This division is driven predominantly by high purity silicon, though there are some other items in the division, including road salts.

3   Estimates by PHOTON Consulting in its data platform, *The Wall*, based on a dedicated team of analysts analyzing more than 1,000 solar power companies.

4   Data from the annual report of each company.

5   PHOTON Consulting *Silicon Monthly* as of February 2011 and PHOTON Consulting's *The Wall* data platform, http://www.photonconsulting.com/consulting_services.php.

6   PHOTON Consulting's *The Wall* data platform; Global weighted average price of $55/kg with global weighted average all-in cost of $37/kg equates to operating profit of $18/kg and a 33% operating margin ($18/kg operating profit ÷$55/kg price).

7   There were many large bankruptcy filings in 2008 and 2009. In 2009, major bankruptcies included Lyondell ($29 billion in assets when the filed in January 2009), GGP ($30 billion in April 2009), Chrysler ($39 billion in April 2009), Thornburg Mortgage ($37 billion in May 2009), GM ($91 billion in June 2009) and CIT ($70 billion in November 2009). These are just examples. With each of these bankruptcies came many others. For more information, see Christopher Tkaczyk, "The 10 Largest U.S. Bankruptcies," *CNN Money*, Nov. 1, 2009, http://money.cnn.com/galleries/2009/fortune/0905/gallery.largest_bankruptcies.fortune/ and Anders Ross, "22 Largest Bankruptcies in World History," *InstantShift*, Feb. 3, 2010, http://www.instantshift.com/2010/02/03/22-largest-bankruptcies-in-world-history/.

8   PHOTON Consulting's *The Wall* data platform.

9   James Paton, "Origin Says Solar Business to Expand after Fivefold Sales Gain," *Bloomberg*, Feb. 24, 2011, http://www.bloomberg.com/

news/2011-02-25/origin-says-australian-solar-business-to-expand-after-fivefold-sales-gain.html.

10   *Solar Annual 2010–2011: Cash In*, http://www.photonconsulting.com/solarannual2010-2011/

11   All data are rough estimates. US-dollar estimates based on exchange rates from QANDA™.

12   For additional information, please see "Solar Annual 2009: Total Eclipse," http://www.photonconsulting.com/solar_annual_total_eclipse.php; "Solar Annual 2010–2011: Cash In", http://www.photonconsulting.com/solarannual2010-2011/; and PHOTON Consulting's data platform, *The Wall*; To put SMA's market share in perspective, the world's largest maker of silicon (Hemlock) has only 20% market share and the world's largest maker of solar power modules (Suntech) has only 9% market share.

13   SMA's *2010 Annual Report* provides an overview of historical data in various places throughout the report, http://www.sma.de/fileadmin/fm-dam/Investor_Relations/Downloads/Finanzberichte/2010/2011-03-30_SMA-Annual_Report_2010-UK_web.pdf.

14   "Cypress Semiconductor Corporation, " Hoovers, http://www.hoovers.com/company/Cypress_Semiconductor_Corporation/rfctxi-1.html.

15   Ibid.

16   For a case study on some of SunPower's history, see the case study by Rebecca Henderson, Joel Conkling and Scott Roberts, *SunPower: Focused on the Future of Solar Power* (MIT Sloan Management, 2007), https://mit-sloan.mit.edu/MSTIR/sustainability/SunPower/Documents/07-042-Sun-Power-Henderson.pdf.

17   "MEMC Electronic Materials, Inc," Hoovers, http://www.hoovers.com/company/MEMC_Electronic_Materials_Inc/ccftri-1.html.

18   Ibid.

19   "Sharp Corporation," Hoovers, http://www.hoovers.com/company/Sharp_Corporation/crkxhi-1.html.

CHAPTER 5

1   During the next 1,000 days, answering the questions, "Where is value?" and "How to capture it?" will likely not be technology driven, but will likely be market-insight driven. Although this may be a disappointment to members of the MIT community that cheer for technology as a differentiator based on faster speed and/or lower cost, the dynamic in industries involved with negative network effects is that the fastest and most thorough understanding of the market is a major enabler of "winning." Also, it is important not to confuse "technology differentiation" with "cost leadership." In the next 2–3 years, leading companies will need to establish cost leadership if they are going to pursue successful business models. However, I am convinced that this cost leadership will emerge over time as the result of knowing the market well and, as a result, being able to make very clear decisions about procurement, capacity, skills and so on in order to bring down the cost curve quickly. My point is that I do not believe that this effort to drive down costs will come from technology as much as from market insights.

2   Julie Gallagher, "GMA Merchandising Conference; Distributers Urge Focus on Demand Driven Supply," *Supermarket News*, October 2, 2006, http://business.highbeam.com/4524/article-1G1-152339500/gma-merchandising-conference-distributors-urge-focus.

3   "Farmers Markets and Local Food Marketing," Agricultural Marketing Service of the U.S. Department of Agriculture, http://www.ams.usda.gov/AMSv1.0/farmersmarkets.

4   "Grocery Stores & Supermarkets," Hoovers, www.hoovers.com/industry/grocery-retail/1535-1.html.

5   All data are rough estimates. Rural-urban continuum code 1–9 represents a spectrum of urban [lower number starting at (1)] to rural [higher number rising to (9)] geographic locations of specific farmers markets.

6   All data are rough estimates. These data are based on interviews with customers who self-report in response to the survey question. One piece

of data suggesting that these self-reported estimates may be accurate is arrived at by checking average weekly sales estimates from self-reported data ($31.27 per customer per week in 2010) compared to terminal data aggregated for credit card customers ($27.18 per customer per week in 2010 not including any cash purchases by those customers).

CHAPTER 6

1   Michael Rogol, *Why Did the Solar Power Sector Develop Quickly in Japan?* (MIT Engineering Systems Division, 2007).

2   PHOTON Consulting's data platform, The Wall.

3   *First Solar Annual Report* for 2007, http://library.corporate-ir.net/ library/20/201/201491/items/303509/2007AR.pdf; *First Solar Annual Report* for 2010, http://phx.corporate-ir.net/External.File?item=UGFy ZW50SUQ9NDI1NzMyfENoaWxkSUQ9NDQxMDY1fFR5cGU9MQ= =&t=1.

4   Annual reports from 2007, 2008, 2009 & 2010 of the respective companies.

INDEX

A

Accenture, 97

action map, building a 1,000-day, 150–53, 154–55

adoption patterns. *See* customers' adoption patterns

AEG, 88

Agricultural Marketing Services (USDA), 111–12

Angelou, Maya, 161

Arthur D. Little (ADL), 97

Australia, solar power industry growth in, 79

B

banks and online bill pay customers, 15–16

Bicknell, Thomas William, 178n3

Boston Consulting Group, 19

Boston, Massachusetts, 41–42, 106–8

Britt, Frank, 96–100, 180n7

growth in new farmers markets, 100–101, 109–10, 111–14

in Massachusetts, 106–8, 110–16

negative network effect analysis, 101–6

profit impact on centralized food, 104–6

spending by customer, 116

trend augmentation, 116–17

financial challenge as negative network effect, 23, 40–41, 52. *See also* negative
   network effects impacting financial markets

financial model of industry sector. *See* sector accounting

First Solar, 153

fixed cost structure

importance of volume growth, 51–53, 62

mobile phones, 49

overview, 4–5, 48

USPS, 17, 20, 23, 172*n*13

food infrastructure

distributed foods, 57, 58–60, 62–63, 117

and farmers markets, 103, 104–6

loss of highest margin demand, 117

plausible negative network effects, 55, 57–61, 156

See also farmers markets

food trucks, 60, 117

Food & Wine magazine, 60

Forrester, 19

France, solar power industry growth in, 35–36

G

GCL, 70–71, 72

General Accounting Office (GAO), 23

geographic region density and farmers markets, 111–14

Germany, solar power industry growth in, 32–34, 39, 40, 55–56, 140, 174n14

global weighted average price, 133–34

"going local" mega-trend in grocery retailing, 102

government intervention in solar power development, 31

Greece, solar power industry growth in, 35–36

grid price assumptions as key uncertainty in a sector, 146

grow-at-home food kits, 57, 58–60, 62–63

grow-at-home foods, 117

growth. *See* volume growth

H

Harris Interactive, 12–13

Hemlock, 70–71, 72

high-end customers. *See* wealthy customers

high price products, 114

History of the State of Rhode Island and Providence Plantations, The (Bicknell),
     178n3

Hoovers, 86, 87

I

IBM, 98

income statements, 128–29, 133, 136–37, 157

industries affected by energy network shifts, 7

infrastructures

    communications, 55, 156

    telecommunications, 20, 49, 51, 56–57, 179n6

    water, 55, 117, 156

    See also distributed infrastructures; energy infrastructure; food infrastructure;
        traditional centralized infrastructures; traditional infrastructure networks

"Projecting U.S. Mail Volumes to 2020" (Boston Consulting Group), 19

R

REC, 70–71, 72

Recurrent Energy, 87–88

revenue and sector accounting, 133–34

risk and reward, 149–50

risk identification, 146–47

Rockefeller, John D., 7–8, 128, 157

Rogol, Susan, 57–59, 110

Roosevelt, Theodore, 121

S

scale

    competing on superior analytics and, 98

    and negative network effect, 22, 23–24, 40, 41

    and speed, of technology transitions, 23–24, 41, 84–85

scenarios with multiple business models, 145–50, 154

seasonal factors and solar power, 33–34

sector accounting

    cost, 134–36, 144

    demand curve, 129–30, 131–34

    finishing touches, 137–39

    income statements, 128–29, 133, 136–37

    overview, 128

    price, 133–34

    price and volume adoption patterns along supply chain, 139–41

    profit, 129, 135, 136–37

    revenue, 133–34

V

www.ingramcontent.com/pod-product-compliance
Lightning Source LLC
Chambersburg PA
CBHW031404180326
41458CB00043B/6613/J